Future of Business and Finance

The Future of Business and Finance book series features professional works aimed at defining, analyzing, and charting the future trends in these fields. The focus is mainly on strategic directions, technological advances, challenges and solutions which may affect the way we do business tomorrow, including the future of sustainability and governance practices. Mainly written by practitioners, consultants and academic thinkers, the books are intended to spark and inform further discussions and developments.

Heike Proff · Pedro J. Marrón ·
Elmar Pritsch · Harald V. Proff
Editors

Future of the Automotive Industry

From System Integration to Ecosystem Play

Springer

Editors
Heike Proff
Allgemeine Betriebswirtschaftslehre &
Internationales Automobilmanagement
University of Duisburg-Essen
Duisburg/Essen, Germany

Elmar Pritsch
Deloitte Consulting Germany
Düsseldorf, Germany

Pedro J. Marrón
Networked Embedded Systems Group
University of Duisburg-Essen
Duisburg/Essen, Germany

Harald V. Proff
Deloitte Consulting Germany
Düsseldorf, Germany

ISSN 2662-2467 ISSN 2662-2475 (electronic)
Future of Business and Finance
ISBN 978-3-032-26307-0 ISBN 978-3-032-26308-7 (eBook)
https://doi.org/10.1007/978-3-032-26308-7

This Springer imprint is published by the registered company Springer Nature Switzerland AG
The registered company address is: Gewerbestrasse 11, 6330 Cham, Switzerland

Preface

For a long time in the automotive industry, it was system integration that promised profits for traditional manufacturers and large suppliers with a strong focus on the static milestone start of production (SOP). Today, however, system integration is increasingly being replaced by interactions in ecosystems, working together to realize a central value proposition – either on the basis of joint innovation in an innovation ecosystem or through reuse of common assets in a transaction ecosystem. Both ecosystem approaches allow to continuously improve and enhance the functions and service features dynamically during the entire product lifecycle – even after SOP. Traditional automotive companies must therefore play the ecosystem game, in order to remain competitive and attractive to the capital markets. However, there are major challenges in this endeavour. In innovation ecosystems, these are often of a technical nature, such as the transition to a Software-Defined Vehicle (SDV) with a central electric control unit that can be updated over-the-air (OTA). Transaction ecosystems, on the other hand, are often not economically profitable, such as Mobility as a Service (MaaS) ecosystems, which bundle various mobility services such as car and bike sharing, public transport and taxis in a single platform across companies, supplemented e.g. by billing services.

We are dealing with this paradigm shift from system integration to ecosystem play in the automotive industry, both scientifically – from a technical perspective at LocosLab GmbH and from the Computer Science chair for Networked Embedded Systems and from an economic perspective from the Chair of General Business Administration & International Automotive Management at the University of Duisburg-Essen in Germany – as well as practically in Deloitte's Global Automotive network. The result is this book: "Future of the Auto Industry – From system integration to ecosystem play".

The book takes a fundamental look at the increasing cooperation between automotive companies in innovation and transaction ecosystems and the technical architecture of ecosystems (customer application, middleware and backend). On this basis, it first looks at SDV innovation ecosystems – future challenges of homologation and opportunities for business development as well as the role of traditional original equipment manufacturers (OEM)s. Subsequently, the status quo of MaaS transaction ecosystems worldwide will be presented and the future coexistence of MaaS and private car ownership as well as the role of traditional OEMs will be discussed.

We would like to thank the teams and the Chairs of ABWL & International Automotive Management and Networked Embedded Systems, LocosLab GmbH as well as Deloitte and we hope that the book will find many readers. We also hope that the book will provide inspiration, and appeal to both academics and practitioners and encourage collaboration between different disciplines.

Duisburg, Germany — Heike Proff
Essen, Germany — Pedro J. Marrón
Düsseldorf, Germany — Elmar Pritsch
Düsseldorf, Germany — Harald V. Proff
June 2026

Contents

Editors and Contributors

About the Editors

Prof. Dr. Heike Proff Studied business administration in Frankfurt and Mannheim; Ph.D. from the University of Frankfurt; habilitation from the University of in Mannheim; conducted research in Japan, Ghana, Korea, Brazil, and the United States. From 2004 to 2009, held the Zeppelin Chair in International Management at Zeppelin University in Friedrichshafen. Since 2009, Chair of Business Administration & International Automotive Management at the University of Duisburg-Essen and coordinator of the master's program "Automotive Engineering & Mobility Management". Research focuses on strategic and international management, particularly in the automotive industry, with empirical studies primarily on the transformation of the automotive industry; member of the "Program on Vehicle and Mobility Innovations (PVMI, formerly IMVP)" and "The International Network of the Automobile (Gerpisa)."

Prof. Dr. Pedro J. Marrón B.S. and M.S. in Computer Engineering from the University of Michigan, Ann Arbor, USA; Ph.D. from the University of Freiburg; habilitation from the University of Stuttgart. From 2007 to 2009, held the Chair of Networked Embedded Systems at the University of Bonn. Since 2009, has held the Chair of Networked Embedded Systems at the University of Duisburg-Essen. Since 2020, Chairman of the Board of GUIDE, the startup and entrepreneurship center at the University of Duisburg-Essen. Since 2022, Vice Rector for Transfer, Innovation, and Digitalization at the University of Duisburg-Essen. Research focuses on IoT, smart cities (especially smart mobility), robotics, and the networking of embedded systems.

Dr. Elmar Pritsch is a Partner at Deloitte Consulting and advises worldwide leading companies in the automotive industry in the context of software defined vehicles. He focuses on strategy development and implementation based on his experiences with digital transformation, emerging ecosystems, alliance frameworks and partnerships.

After studying computer science with a minor in business administration and subsequent doctorate at the RWTH Aachen University, he worked as a strategy consultant and manager for international corporations in the telecommunications, IT services and automotive sectors. Dr. Elmar Pritsch has more than 15 years of experience as a board and supervisory board member and earned his MBA from Edinburgh Business School in the UK.

Dr. Harald V. Proff Studied industrial engineering in Darmstadt. After earning his doctorate in 1993, he worked at the Daimler-Benz Group (product development, manager for industrialization projects); in 2001, he earned his habilitation in economics at the Technical University of Darmstadt. From 2001 to 2012, management consultant at McKinsey & Company Inc. (primarily in operational transformation programs at large mid-sized companies and private equity firms). He has together 20 years of consulting experience in operations (product development, cost reduction, digital operations). He was a senior partner in operations and co-founder of the Deloitte Digital Factory at Deloitte Consulting Germany, and has been Global Sector Lead for Automotive since 2020.

Contributors

Arman Arzani University of Duisburg-Essen, Duisburg/Essen, Germany

Corinna Cruceru-Weisbrod Deloitte Consulting Germany, München, Germany

Marcus Handte University of Duisburg-Essen, Duisburg/Essen, Germany

Luca Husemann University of Duisburg-Essen, Duisburg/Essen, Germany

Dr. Arne Jeppe University of Duisburg-Essen, Duisburg/Essen, Germany

Fabian Langanki University of Duisburg-Essen, Duisburg/Essen, Germany

Prof. Dr. Pedro J. Marrón University of Duisburg-Essen, Duisburg/Essen, Germany

Dr. Elmar Pritsch Deloitte Consulting Germany, Düsseldorf, Germany

Prof. Dr. Heike Proff Allgemeine Betriebswirtschaftslehre & Internationales Automobilmanagement, University of Duisburg-Essen, Duisburg/Essen, Germany

Dr. Harald V. Proff Deloitte, Düsseldorf, Germany

Erik Thorwirth Deloitte Consulting Germany, Düsseldorf, Germany

Nico Zauner Deloitte, Düsseldorf, Germany;
IESE Business School, Barcelona, Spain

Introduction: From System Integration to Ecosystem Play in view of New Competitors

1

Heike Proff and Harald V. Proff

Abstract

For a long time, the automotive industry was viewed as technologically mature, following an evolutionary and incremental design path. Established global automakers have dominated the industry and played the role of "system integrators", controlling final assembly, bearing ultimate responsibility for product quality and liability, distributing the values and trying to become bottlenecks. Thereby, systems integration was seen as a unique competence of automotive companies which protects them in the competition and ensured its dominance in the automotive value chain. In consequence, the car manufacturers dominate as system integrators that produce all important components. The introduction shows how this system integration has evolved into an ecosystem play.

For a long time, the automotive industry was viewed as technologically mature, following an evolutionary and incremental design path [1]. Established global automakers have dominated the industry and played the role of "system integrators", controlling final assembly, bearing ultimate responsibility for product quality and liability [2], distributing the values and trying to become bottlenecks. Thereby, systems integration was seen as a unique competence of automotive companies

H. Proff (✉)
Allgemeine Betriebswirtschaftslehre & Internationales Automobilmanagement, University of Duisburg-Essen, Duisburg/Essen, Germany
e-mail: heike.proff@uni-due.de

H. V. Proff
Deloitte, Düsseldorf, Germany
e-mail: hproff@deloitte.de

H. Proff et al. (eds.), *Future of the Automotive Industry*, Future of Business and Finance, https://doi.org/10.1007/978-3-032-26308-7_1

which protects them in the competition and ensured its dominance in the automotive value chain [3, 4]. In consequence, the car manufacturers dominate as system integrators that produce all important components.

The importance of system integration competence decreases somewhat in the transition to electric vehicles (EVs), but still remains important. EVs mean a "disruption through substitutes" [2], but also need an integrated product architecture, although they are easier to build with fewer parts. Therefore, the most likely scenario concerning EVs is that the system integrator role will still be essential in the future. The prediction is that hardware/software integration will become the most valued capability for either incumbents or new entrants – and this capability will require a combination of digital technology and automotive expertise. This means that pure contract manufacturing is not possible for auto OEMs [5]. Therefore, there is a significant difference to computer manufacturers, for example, who have lost their dominance. While in the computer industry plug and play dominates, i.e. suppliers provide operating systems and software whereas OEMs only assemble; the automotive industry is more complex. OEMs have complete vertical-integration competence (even for EVs).

The importance of system integration capabilities has only decreased significantly in the course of digitalization and with a shift in profit pools in the automotive and emerging mobility industry towards.

- on the one hand, tech companies ("brain suppliers") like electronic and software providers that use digitalization on a product-related basis and are moving towards connected, highly automated or autonomous electric cars with over-the-air updates in Software-Defined Vehicles as well as
- on the other hand, mobility service providers (customer value player) like (ride- and car-)sharing companies that use digitalization on a customer-related basis and are moving towards shared, connected and mostly electric mobility [2, 6].

Both are competing on software architecture and data platforms in the automotive industry and have triggered a substantial value shift towards new entrants in the auto industry, that bring critical components of the new automotive value chain (sensors, IT hardware, communication, control software, and platform operating system). This means that production is no longer a central asset that has to be done in-house and the bottleneck is changing. Through digitalization, the threat scenario is derived, that tech and sharing companies drive the "CASE-revolution" with CASE as "connected, autonomous, shared, electric"[1] and push back or even displace established automotive companies [2].

Since early 2024, however, the CASE "revolution" seems to be slowing down in many countries which makes the threat scenario less threatening [6]. At the latest since the war in the Ukraine and the rise of nationalist aspirations in many countries, the "new normal" is already being called into question again. Hybrid drives, hydrogen and e-fuels are being propagated, a breakthrough for fully autonomous driving, especially in everyday mixed traffic, is not in sight and networked, shared mobility with mobility cards and mobility budgets is not yet paying off – partly because customers want it but do not use it – e.g. because people's mobility behaviour is highly habitualized (i.e. they hold on to their own vehicle with an internal combustion engine). "There is no CASE" is the (premature) conclusion and it is assumed that the established system integrators will continue to dominate the automotive industry.[2]

Even though these challenges exist and it is important that the incumbents seize global market opportunities with combustion engines and plug-ins that have higher margins, as all (global) profit potential must be tapped for the extremely expensive CASE transformation, the CASE trend has slowed down, but has not been broken. This is supported by the fact that the CASE transformation follows from the global megatrends of urbanization, decarbonization, digitalization and demographic change, which continue unabated. In line with CO_2 ceilings in regulations such as the Paris Climate Agreement, its implementation in the European Green Deal[3] and in commitments by many countries and regions to a net-zero climate target by 2050 or 2060 at the latest, many automotive companies increasingly see electromobility as "the new normal" [7], have made efforts to develop sharing concepts and accept the trend towards connectivity and highly automated driving in Software-Defined Vehicles with new vehicle architecture. Above all, however, new technology company competitors from the USA and China, and also ride-sharing providers are driving the change [6].

This means, that system integrators will become less important with new competitors entering the markets (brain suppliers and sharing providers, and also new competitors from the USA and China), but neither the extreme scenario of there being no CASE and automotive companies continuing to act as system integrators nor the other extreme scenario of CASE revolutionizing the automotive and mobility markets and tech and mobility companies completely displacing automotive companies is realistic. This raises the question of what role the established car manufacturers and major automotive suppliers will realistically play in the future, i.e. how much of their former dominance they will be able to retain.

Dominant companies are those that can determine the distribution of value creation in an industry. Up to now, established car manufacturers and suppliers that "manufacture motor vehicles and parts" have had the largest share of value added (over 80%). Further value added is primarily generated by the automotive trade and the aftermarket ("maintenance and repair of motor vehicles" and "trade in motor vehicle parts and accessories"), see the statistical classification of economic activities in the European Community [8]. Connected and highly automated (in future autonomous), shared and electrified mobility, however, opens up new value creation channels for start-ups and companies from outside the industry, e.g. from brain supplier and sharing providers [2, 9]. In China, for example, there are already takeovers of vehicle manufacturers by telecommunications companies. New automotive companies from North America and China are taking further shares away from the already declining value creation of established car manufacturers and suppliers. In China alone, there were 95 (!) different car brands in spring 2024, where every second vehicle sold is now electrified. Even if these new competitors must consolidate into fewer than 10 large groups and (want to) compete in the premium segment, they can achieve a more favourable cost position as the "new kids on the block" without the burden of existing production structures.

With regard to future value creation and thus the importance of established automotive companies, we conducted a survey of 250 global automotive companies in the most important markets in 2023 ([10] and related [6]). We wanted to know whether they expect a significant change in the automotive value creation structure by 2030 and 2035. The result is clear: They expect a radical change by 2030, as they assume that the established car manufacturers, suppliers and dealers will then only have a 40% share of automotive value creation, while 60% will be accounted for by new players in the automotive industry, primarily Chinese and those from outside the industry. After that, however, they assume that the value creation structures will stabilize and expect almost no further changes until 2035 (Fig. 1.1a).

The survey participants, however, also expect shifts within the value creation of established automotive companies. If they still generate almost half of their value added with hardware in 2023, they think that in 2030 and 2035 only a third will be generated with hardware and around a further third each with the electronics and software and sharing business, with electronics and software becoming slightly more important (Fig. 1.1b).

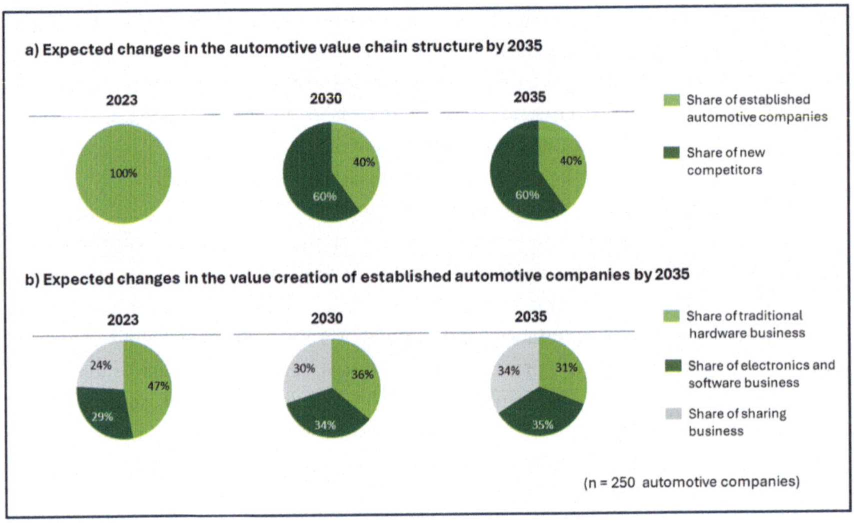

Fig. 1.1 Expected changes in value creation (Deloitte 2024 [10] and related Proff 2025 [6])

The future competitive landscape, however, depends on which economic model prevails:

- either the model of the established automotive companies, which continuously increases their productivity and, as a result, their profitability in order to satisfy their shareholders
- or the model of the new competitors, who rely on market share to drive up their stock market value ([11] and related [6]).

This is because the models hinder each other—if the stock market value of the new competitors increases, there is a risk that they will take over the established companies, unless they are state or family-owned. Most automotive companies are worth little from the perspective of the large technology companies. However, if the incumbents create an attractive offering based on productivity gains, even the stock market hype surrounding new providers can collapse if they are unable to keep up and the stock market loses confidence due to declining innovative strength.

Therefore, the established car manufacturers can only survive if they increase the speed of their transformation and use their profitability to catch up quickly with modern technologies. To do this, however, they need attractive offers instead of losing themselves in back end actions, such as slowing down the transition to electromobility.

To speed up, in the long-term, far-reaching and disruptive CASE revolution, established automotive companies need high investments in connectivity, autonomy, sharing and electrification (CASE technologies) that are required in order to compete with technology companies and are demanded by capital markets [12]. New skills, e.g. in battery technology and artificial intelligence are needed.

They also need transformation skills to adapt their operational capabilities to the radical technological changes [6]. Such transformation capabilities are (1) capabilities of operational agility that enable binding decision-making processes based on minimum viable solutions [13], (2) ambidextrous management skills of the old and the new business in transformation [14], (3) collaboration skills in novel ecosystems [15].

Ad 1: Operational agility capabilities enable binding decision-making processes based on minimally viable solutions that are repeatedly reviewed, adapted and ultimately improved in several loops. In the capital-intensive automotive industry, however, hybrid agility is required [16]. This is because, in addition to agility, an efficient orientation through economies of scale and scope is important.

Ad 2: Ambidexterity is the capability to reduce value creation in the old (analog and combustion-based) business and simultaneously increase it in the new innovative (digital and electromobile) business [14], as capital allocation must be fixed within the value creation network. However, there is a conflict of objectives between agility and efficiency [13], as companies lose agility (flexibility) by fixing the use of capital (efficiency). It is a paradox because agility capabilities and the optimization of economies of scale and scope are mutually exclusive.

Ad 3: The capability to join together in ecosystems means more than a basic capability to cooperate or to work together in supply chains [17]. It is a special capability to participate in an "alignment structure of the multilateral set of partners that needs to interact in order for a focal value proposition to materialize" [15], mostly based on digital platforms [18]. These ecosystems have gained increasing attention, both, in academia and in practice [19, 20] and they aim at joint value creation and individual value creation of the partners [21–23]. It is important to distinguish between innovation ecosystems [24, 25] that offer a venue for joint innovation via innovation platforms and transaction ecosystems [26] that are multi-sided marketplaces enabling the reuse of "common resources" (activities and assets) of the ecosystem partners via transaction platforms [27].

Especially the capability to participate in ecosystems can help traditional automobile companies gain and further develop CASE-competences and also fulfil the demands of capital markets for asset-light strategies requiring (low) investments in traditional technologies at the same time [27]. This is because ecosystems make "business scalable flexibly without investments in heavy assets" [28]. Therefore, for established automotive companies, both types of ecosystems are important;

- innovation ecosystems, when working together with electronic and software companies to create e.g. mapping solutions, tracking, positioning and other forms of connected and high automated mobility with precise, dynamic data and
- transaction ecosystems, when participating in platforms for MaaS, which bundle various mobility services such as car and bike sharing, public transport and taxis into a single platform across companies, supplemented by billing services [29].

Ecosystems mean a fundamental "paradigm shift" in the strategies of national and international companies from exclusiveness and high capital intensity to openness and an asset light leverage of partners' assets and knowledge [27, 28]. In the capital-intensive automotive industry, this results in the shift from system integration to ecosystem play in view of new competitors – brain suppliers as well as mobility service providers that threaten and are already reducing their dominance in the automotive and mobility industry.

After the opportunities of ecosystems were initially used, primarily by technology companies, to increase their technology share [12], they are now also being recognized by manufacturing companies [30], especially automotive companies [2, 31]. However, capital-intensive automotive companies have problems with the development of ecosystems. They often seem unfeasible [32], mainly due to the risk of mutual dependencies between the partners [33]. In addition, there is the threat of a capability drain to partners [34] who are also competitors outside the ecosystem (coopetition, [35]), and there is a risk of misjudgment, mistrust and incorrect timing, as well as insufficient cybersecurity and regulatory complexity [36]. Therefore, it is important for established automotive companies to develop the ability to play the ecosystem game. Only then, they can also properly exploit the opportunities of digitalization for a "digital expansion" of their business models in the ecosystem [37], such as a cooperation with Google and Apple in the area of navigation systems, without being overtaken by the new complementary players in the digital game.

Our book deals with the future of the automotive industry, which is moving from system integration to the ecosystem game. On the one hand, we want to examine the question of where traditional automotive companies will remain in the future ecosystem game. On the other hand, despite the heterogeneity of the individual companies and national markets, we also want to point out lines in the complexity of industry change and encourage further thought and action.

To this end, we will first examine the "Increasing coopetition between automotive companies in innovation and transaction ecosystems" (Chap. 2) and the "Technical architecture of ecosystems: Customer application, middleware and backend" (Chap. 3). The coopetition between traditional automotive companies and technology companies specifically in innovation ecosystems is then examined: "Connected and highly automated driving innovation ecosystems worldwide" (Chap. 4) and, as a deep dive, the "Development of a software-defined vehicle architecture" (Chap. 5). This is followed by an examination of "Mobility as a Service transaction ecosystems worldwide" in Chap. 6, first the status quo of these ecosystems and approaches to improving them, then in Chap. 7 "Future coexistence of Mobility as a Service and private car ownership" with the question of where the traditional automotive companies will remain in the MaaS game in the future. The book ends with an "Outlook: Future of the Auto industry".

Notes

1. https://group.mercedes-benz.com/innovation/case.html (retrieved on 31.5.2024).
2. E.g. the discussion at the 32nd International Colloquium of Gerpisa (https://gerpisa.org, retrieved on 31.5.2024).
3. European Commission (2019): The European Green Deal. COM (2019) 640 final. Brussels.

References

1. Abernathy, W. J., & Utterback, J. M. (1978). Patterns of industrial innovation. *Technology Review, 80*(7), 40–47.
2. Adner, R., & Lieberman, M. (2021). Disruption through complements. In: Strategy. *Science, 6*(1), 91–109.
3. Jacobides, M. G., MacDuffie, J. P., & Tae, C. J. (2016). Agency, structure, and the dominance of OEMs: Change and stability in the automotive sector. *Strategic Management Journal, 37*(9), 1942–1967.
4. MacDuffie, J. P. (2018). Response to Perkins and Murmann: Pay attention to what is and isn't unique about Tesla. *Management and Organization Review, 14*(3), 481–489.

5. Perkins, G., & Murmann, J. P. (2018). What does the success of Tesla mean for the future dynamics in the global automobile sector? *Management and Organization Review, 14*(3), 471–480.
6. Proff, H. (2025). Strategic endgames – Wie sieht die Anbieterlandschaft der Zukunft aus? In H. Proff (Ed.), *New players in mobility: Technische und betriebswirtschaftliche Aspekte*. Springer Gabler.
7. Proff, H. (Ed.). (2023). *Towards the new normal in mobility: Technische und betriebswirtschaftliche Aspekte*. Springer Gabler.
8. Eurostat. (2008). NACE Rev. 2: *Statistische Systematik der Wirtschaftszweige in der Europäischen Gemeinschaft*. https://ec.europa.eu/eurostat/documents/3859598/5902453/KS-RA-07-015-DE.PDF
9. IPE Institut für Politikevaluation, fka, & Roland Berger. (2019). *Automobile Wertschöpfung 2030/50: Eine Studie im Auftrag des Bundesministeriums für Wirtschaft und Energie. Endbericht*. Bundesministerium für Wirtschaft und Energie. bmwi.de.
10. Deloitte. (2024). *Status of the transformation in the automotive industry*. Forthcoming.
11. Proff, H., & Proff, H. V. (2013). *Dynamisches Automobilmanagement: Strategien für international tätige Automobilunternehmen im Übergang in die Elektromobilität* (2. Aufl.). Springer Gabler.
12. Ceccagnoli, M., Higgins, M. J., & Kang, H. D. (2018). Corporate venture capital as a real option in the markets for new technology. *Strategic Management Journal, 39*(13), 3355–3381.
13. Teece, D. J., Peteraf, M., & Leih, S. (2016). Dynamic capabilities and organizational agility: Risk, uncertainty, and strategy in the innovation economy. *California Management Review, 58*(4), 13–35.
14. Raisch, S., & Tushman, M. L. (2016). Growing new corporate businesses: From invitation to graduation. *Organization Science, 27*(5), 1237–1257.
15. Adner, R. (2017). Ecosystem as structure: An actionable construct for strategy. *Journal of Management, 43*(1), 39–58.
16. Proff, H. (2019). *Multinationale Automobilunternehmen in Zeiten des Umbruchs: Herausforderungen - Geschäftsmodelle - Steuerung*. SpringerGabler.
17. Kapoor, R. (2018). Ecosystems: Broadening the locus of value creation. *Journal of Organization Design, 7*(1), 1–16.
18. Gawer, A. (2022). Digital platforms and ecosystems: Remarks on the dominant organizational forms of the digital age. *Innovation: Organization & Management, 24*(1), 110–124.
19. Jacobides, M. G., Cennamo, C., & Gawer, A. (2018). Towards a theory of ecosystems. *Strategic Management Journal, 39*(8), 2255–2276.
20. Adner, R. (2021). *Winning the right game: How to disrupt, defend, and deliver in a changing world*. MIT Press.
21. Brandenburger, A. M., & Stuart, H. W. (2007). Biform games. *Management Science, 53*(4), 537–549.
22. Dyer, J. H., Singh, H., & Hesterly, W. S. (2018). The relational view revisited: A dynamic perspective on value creation and value capture. *Strategic Management Journal, 39*(12), 3140–3162.
23. John, K., & Ross, D. G. (2022). How a firm's value capture affects value creation in its ecosystem. *Academy of Management Review, 47*(4), 646–667.
24. Ketonen-Oksi, S., & Valkokari, K. (2019). Innovation ecosystems as structures for value co-creation. *Technology Innovation Management Review, 9*(2), 25–35.
25. Wulfert, T. (2023). Boundary resource management in innovation ecosystems: The case of e-commerce. *Electronic Markets, 33*(1), Article 27.
26. Rochet, J.-C., & Tirole, J. (2003). Platform competition in two-sided markets. *Journal of the European Economic Association, 1*(4), 990–1029.
27. Cusumano, M. A., Gawer, A., & Yoffie, D. B. (2019). *The Business of platforms: Strategy in the age of digital competition, innovation, and power*. HarperCollins Publishers.

28. Cha, H. (2020). A paradigm shift in the global strategy of MNEs towards business ecosystems: A research agenda for new theory development. *Journal of International Management, 26*(3), Article 100755.
29. Karlsson, I., Mukhtar-Landgren, D., Smith, G., Koglin, T., Kronsell, A., Lund, E., Sarasini, S., & Sochor, J. (2020). Development and implementation of Mobility-as-a-Service – A qualitative study of barriers and enabling factors. *Transportation Research Part A: Policy and Practice, 131,* 283–295.
30. Sjödin, D., Parida, V., & Kohtamäki, M. (2022). How large manufacturers digitalize their business models: A framework for orchestrating industrial ecosystems. *California Management Review, 64*(3), 122–144.
31. Kraus, L., & Proff, H. (2024). Tech play and global play in the automotive industry: Mediating goal conflicts with platform-based ecosystems. *International Journal of Automotive Technology and Management, 24*(1), 1–26.
32. Reeves, M., Lotan, H., Legrand, J., & Jacobides, M. G. (2019). How business ecosystems rise (and often fail). *MIT Sloan Management Review, 60*(4), 1–6.
33. Adner, R., & Feiler, F. (2019). Interdependence, perception, and investment choices: An experimental approach to decision making in innovation ecosystems. *Organization Science, 30*(1), 109–125.
34. Krylova, K. O., Vera, D., & Crossan, M. (2016). Knowledge transfer in knowledge intensive organizations: The crucial role of improvisation in transferring and protecting knowledge. *Journal of Knowledge Management, 20*(5), 1045–1064.
35. Bengtsson, M., & Raza-Ullah, T. (2016). A systematic review of research on coopetition: Toward a multilevel understanding. *Industrial Marketing Management, 57,* 23–39.
36. Luo, Y. (2021). A general framework of digitalization risks in international business. *Journal of International Business Studies, 53,* 344–361.
37. Reischauer, G., Engelmann, A., Gawer, A., & Hofmann, W. H. (2024). The slipstream strategy: How product-status OEMs cooperate with platforms to maintain their digital extensions' edge. *Research Policy, 53*(7), Article 105021.

Increasing Coopetition between Automotive Companies in Innovation and Transaction Ecosystems

2

Arne Jeppe, Luca Husemann, and Heike Proff

Contents

Abstract

As the automotive industry is increasingly evolving from system integration to an ecosystem play, they find themselves in an increasing situation of coopetition: They cooperate with partners who are also competitors outside the ecosystem. Coopetition carries the risk of being dependent on the partners or losing skills to them and it can only be motivated by joint value creation with these partners which also offers them the opportunity to participate in the extremely expensive CASE transformation without having to invest too much of their own assets. This chapter examines the governance mechanisms and value drivers that guide coopetition in both innovation and transaction ecosystems.

A. Jeppe · L. Husemann
University of Duisburg-Essen, Duisburg/Essen, Germany
e-mail: arne.jeppe@uni-due.de

L. Husemann
e-mail: luca.husemann@uni-due.de

H. Proff (✉)
Allgemeine Betriebswirtschaftslehre & Internationales Automobilmanagement, University of Duisburg-Essen, Duisburg/Essen, Germany
e-mail: heike.proff@uni-due.de

H. Proff et al. (eds.), *Future of the Automotive Industry*, Future of Business and Finance, https://doi.org/10.1007/978-3-032-26308-7_2

2.1 Introduction

As the automotive industry is increasingly evolving from system integration to an ecosystem play (Chap. 1), they find themselves in an increasing situation of coopetition [1]: They cooperate with partners who are also competitors outside the ecosystem. Coopetition carries the risk of being dependent on the partners [2] or losing skills to them [3] and it can only be motivated by joint value creation with these partners [4] which also offers them the opportunity to participate in the extremely expensive CASE transformation without having to invest too much of their own assets [5].

In order to introduce joint value creation in ecosystems in this chapter, we must first define the term ecosystem (Sect. 2.1). A distinction must be made here between innovation ecosystems and transaction ecosystems. While innovations are created jointly in innovation ecosystems, e.g. in the automotive industry towards SDVs (Chaps. 4 and 5 in this book), services are linked in transaction ecosystems towards the end customers, such as for MaaS (Chaps. 6 and 7 in this book). In Sect. 2.2, we explain the economic significance of ecosystems, which is that they create joint value for the partners that they could not achieve on their own. Finally, we will argue in Sect. 2.3 that different governance mechanisms and value drivers are needed in innovation and transactional ecosystems to jointly create value [6].

2.2 Innovation and Transaction Ecosystems

Particularly in view of the CASE revolution to a connected, autonomous, shared and electric mobility (also Chap. 1), many automotive companies are increasingly integrating their activities [7]. This often results in an "alignment structure of a multilateral set of partners that needs to interact in order for a focal value proposition to materialize" ([8], p. 42). Such alliances are referred to as "structural ecosystems" [8, 9].

The interest in ecosystems has increased exponentially over the last twenty years in both academic and practical literature [10, 11]. Nevertheless, a common, shared understanding of ecosystems only slowly emerged [8, 9, 11]. In the course of research on ecosystems, the originally sociological term [12] was taken up in business studies. There, a "business ecosystem" was initially defined very broadly by Moore [13] as the interdependence and co-evolution of different organizations, even entire industries. Today, however, ecosystems are clearly differentiated from traditional buyer–supplier relationships and the term has been narrowed down to the structural understanding of Adner [8] with (1) binding alignment through complementarity in production, (2) overarching value proposition through complementary services, (3) fixed set of partners with critical resources and (4) multilateral interactions.

The digital transformation [14] and especially the technological modularity of digital platforms [15] allows "hyperspecialization and hyperscaling" [16] in an ecosystem. It enables interdependent components of a system to be produced jointly by different players without major capital investment (asset-light) [17] and strengthens the trend towards ecosystems. Due to the standardization of the interfaces, only limited coordination is required [18].

Thus, there are three key constructs of ecosystems: (1) an overarching value proposition, (2) an operating model of actors with goods/services, data and financial flows between them, and (3) a technical platform (see also [5] and endnote 1). The operating model depicts in an abstract and visual model how a company delivers value to its customers. It is important, because ecosystems are no longer structured and linear like the traditional supplier relationships, but rather very strongly networked relationships with the described technical interfaces between the partners.

The alignment of partners via (often digital) platforms brings "together individuals or organizations so they can innovate or interact in ways not otherwise possible with the potential for nonlinear increase in utility and value" ([14], p. 13]. This also distinguishes between two types of ecosystems:

- innovation ecosystems [19] that offer a venue for joint innovation (e.g. for the development of Software Defined Vehicles) and
- transaction ecosystems [19] as multisided marketplaces that enable the reuse of activities and assets as "common resources" of the ecosystem partners (e.g. for MaaS) via transaction platforms [14].

Ecosystems were initially considered primarily for high-tech companies, because they can significantly increase their technology share via ecosystems with "asset-light" strategies [14], which is rewarded with high capital market valuations. More recently, the importance of such ecosystems for manufacturing companies has also been realized [20], especially in the automotive industry [5, 21]. In particular established or incumbent global automakers, "which have dominated their complex global supply chains" as system integrators "now face challenges in dealing with new partners as the industry moves to an ecosystem structure" ([21], p. 106). For them, regulations and clear mechanisms of value distribution are important, because they have always tried to dominate joint activities (similar to the telecom industry). Today the new structural ecosystems can empower the typically capital-intensive automotive companies in the increasing competition with high-tech companies ("brain suppliers" and sharing companies, [21]) and can enable them to expand their international operations into the growing world markets outside North America, Western Europe and Japan [5].

Coopetition in ecosystems is important in the automotive industry with a global network of large production sites, i.e. in a very capital-intensive industry with a high importance of economies of scale. However, this stands in the way of operational agility, as short, binding and incremental decisions based on minimally viable products represent a "trade-off between agility and efficiency" [22]. Moreover, the high capital intensity brings disadvantages in the increasing competition with high-tech companies [1] and adversely affects the standing of automotive companies in the capital markets [23] as capital intensity in the automotive industry has actually increased by an average of four per cent over the last 20 years (Fig. 2.1). Therefore, “asset-light” strategies are very important in the automotive industry, especially in ecosystems that make “business scalable flexibly without investments in heavy assets” [24].

For automotive companies, both types of ecosystems are conceivable: innovation ecosystems and transaction ecosystems.

An example of an innovation ecosystem that offers a venue for joint innovation via innovation platforms [14] is the platform HERE (https://www.here.com/platform). Through HERE, traditional automotive manufacturers and suppliers are working with software companies to create maps that can be used for mapping solutions, tracking and positioning, and for precise, dynamic data to understand road conditions in real time. Another example is the coopetition necessary to achieve the SDV, which manages its operations, supplements functionality and enables new features (entirely or at least primarily) through software. Coopetition with tech players is important because they will probably have to contribute the central operating system on an open source basis in the future. However, innovation ecosystems towards connected, autonomous and electric mobility are difficult because coopetition between tech players and automotive companies has so far consisted of many announcements. After all, software companies are different from traditional automotive companies. They do not have a zero-defect tolerance, a culture that dominates in traditional OEMs, because software is never finalized. Although software has never had any value for traditional automotive OEMs, they must finally recognize this value.

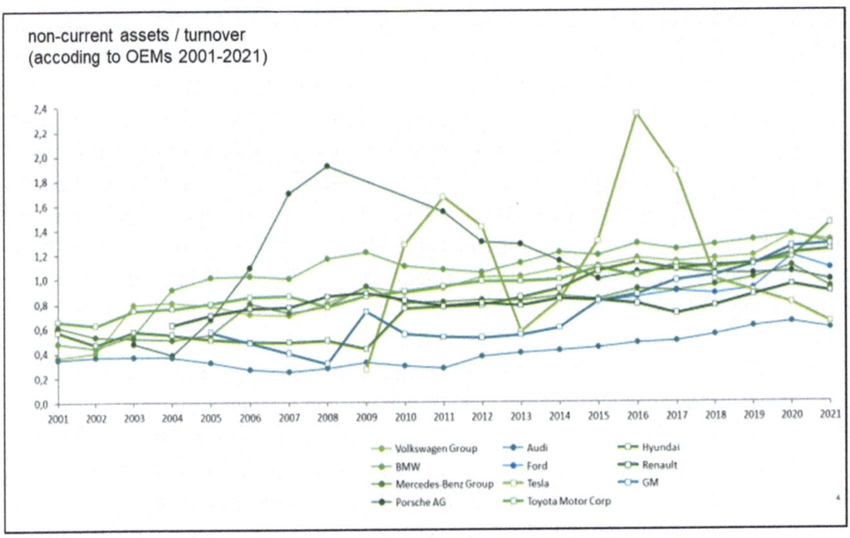

Fig. 2.1 Capital intensity in the automotive industry. *Source* Calculated using annual and financial reports of OEMs (see endnote 1)

In contrast, transaction ecosystems are multisided marketplaces enabling the reuse of “common resources” (activities and assets) of the ecosystem partners via transaction platforms [14, 25]. Examples in the automotive industry are exchange platforms in MaaS ecosystems, which bundle various mobility services such as car and bike sharing, public transport and taxis in a single platform across companies, supplemented e.g. by billing services [26]. This seamless offering of different mobility services [27, 28], is desired by customers [29] but also transport planners because it is essential for achieving the net-zero goals as well as a reduction in traffic, e.g. in city centres [30]. However, these MaaS ecosystems are not (yet) profitable and in practice and most of the many pilot projects do not reach the implementation stage [26]. In March 2024, even the Finnish mobility startup MaaS Global, founded in 2015 as the first provider of a MaaS app (Whim) had to file for insolvency after Covid and the financial crisis.

The difficulties of coopetition in innovation and transaction ecosystems are intensified by the dangers of mispricing, mistrust and mistiming [14], inadequate cybersecurity and regulatory complexity [31]. That is why we need more insights into these ecosystems and their problems in order to understand their opportunities for joint value creation [32], first in the economic significance of ecosystems (the possibility of joint value creation is discussed in the next section) and then in differences in governance mechanisms and value drivers between innovation and transaction ecosystems.

2.3 Economic Significance of Ecosystems

Ecosystems provide potentials for individual companies to create joint value in addition to their own individual value added [33]. From this, the individual company can capture [4] a share of the jointly created value, with the remaining jointly created value still to be negotiated.

Economic “sources of value creation” are derived from the theory of economic rents [34]. For an individual company, they result from the difference between customers’ willingness to pay and supplier’s opportunity costs (forgone benefit of the next best utilization of resources and goods, [35]), see Fig. 2.2a. Negotiations among stakeholders determine the specific shares of value creation [34], with the customer’s share corresponding to the difference between willingness to pay and selling price, the company’s share reflected in the difference between selling price and purchase price, and the supplier’s share determined by the difference between purchase price and opportunity costs.

Ecosystems offer individual companies the potential to create an additional joint value with partners, which they would not have been able to create on their own ([33] and Fig. 2.2b).

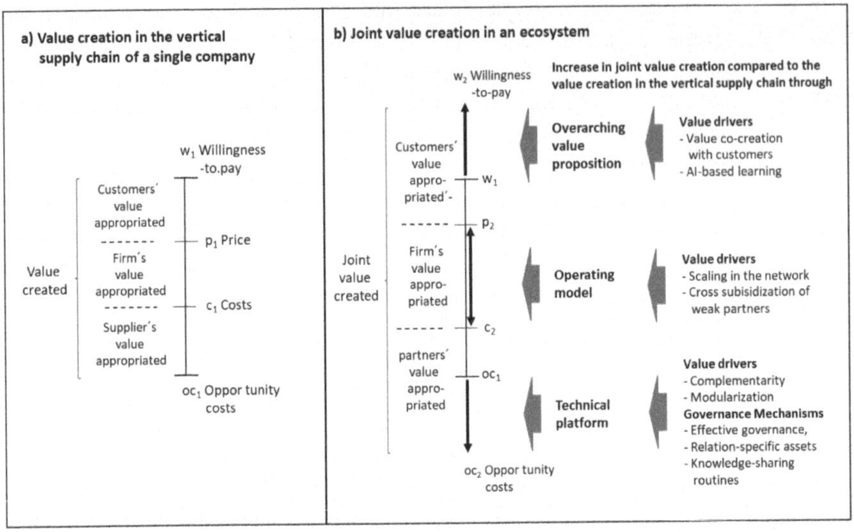

Fig. 2.2 Sources of company's value creation and value capture and joint value creation in an ecosystem. *Source* Own design based on Brandenburger & Stuart [4, 34]

The challenge is, however, that ecosystems often do not create value, but actually destroy value (see Fig. 2.3) because dependencies and interdependencies are no longer controllable by an individual company, but always depend on the interactions of all partners (multilateral relationships, [8]). Especially companies that have long been dominant in traditional manufacturing, often find it difficult to collaborate, as shown by examples in the telecom industry [36], as well as the many failed collaborations, e.g. in the automotive industry and by ecosystems that are not sustainable [14, 37]. These companies often lack flexibility [22] and fear an outflow of resources and capabilities [3] when cooperating with competitors. Moreover, there is a risk of making mistakes in building digital platforms such as mispricing, mistrust, hybris and mistiming [14], as well as in "digital interdependence, information [cyber] security, and regulatory complexity" [31]. Then also no individual value capture succeeds. Therefore, explanations of joint value creation and individual value capture in ecosystems form the core of this kind of coopetition.

The value creation in ecosystems can be justified by extending game-theoretical explanations of value creation and value capture of several partners through biform games [4]. According to the initial explanations, the partner companies involved bring in individual value-added activities autonomously in a non-cooperative phase and negotiate in the cooperative phase about the distribution of the value added [4, 38]. Furthermore, value capture depends on the bargaining power of firms as well as on whether they can capture a bottleneck in the structural ecosystem and thus realize a scarcity premium [39]. However, these explanations need to be expanded (see [5], endnotes 1 and 3):

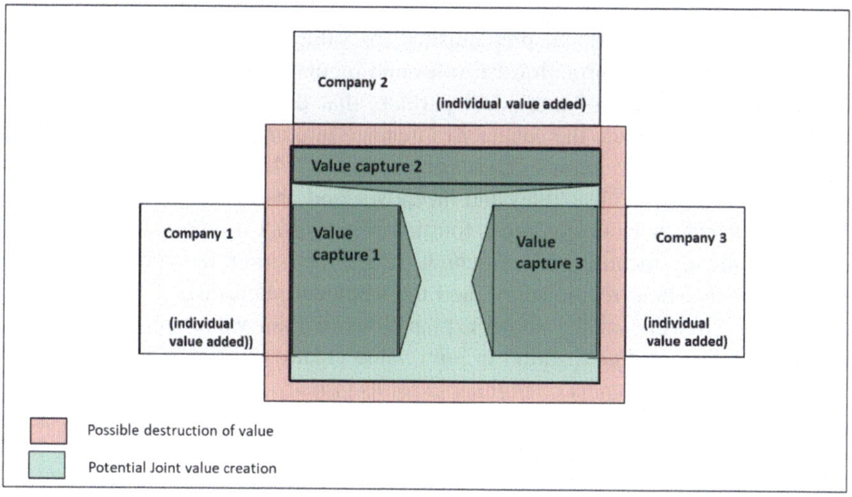

Fig. 2.3 Potential joint value creation, possible destruction of value and individual value capture in ecosystems (based on [4, 37])

A first extension was made by the "service-dominant logic" [40] which takes into account customer interactions and opens the black box "customer" who are neglected in the game theory. However, customers influence the joint value creation by optimizing their "value in use", when putting together an individual, integrated and interactive customer solution. Therefore, the joint value creation and value capture in ecosystems start with the "focal value proposition", which can be derived from customer information (e.g. [8]).

A second differentiation concerns that contrary to the assumption of Brandenburger & Stuart [4], a joint value is not only created by adding the autonomous inputs (value added) of the individual partners in the non-cooperative phase of the competition. In addition, value is created by the cooperation of the partners in the cooperative phase (see endnote 2, [41]). Since this added value cannot be fully accounted for even in the intra-firm resource-based view, the relational view [42] was developed. According to this view, joint value creation and relational rents of partners in value networks depend mainly on complementary resources and capabilities, i.e. supply-side economies of scope [43].

A third extension of the game-theoretic biform games [4] must be made with respect to the distribution of jointly created value and thus with respect to value capture [38]. This is because it cannot be assumed that the distribution of jointly created value, and thus value capture, is purely cooperative through negotiations among partners alone. Instead, according to Adner [8], value capture depends on "activities", i.e. on process activities of "shaping the process of ecosystem development" and on governance activities ([11], p. 2263) of developing rules and regulations [44]. Similarly, the relational view [42] distinguishes three influencing factors as institutional prerequisites for value capture in ecosystems: (1) "effective governance" through trust, rules and regulations [44–46], (2) "interfirm relation-specific assets" of individual partners that determine the role they play in ecosystems and ensure the capacity to manage interorganizational relationships [47], and (3) "interfirm knowledge-sharing routines" [42], above all commitment to knowledge transfer, clear rules and incentives and absorption skills [48]. These three governance mechanisms are a fourth key construct of ecosystems (in addition to the already mentioned overarching value proposition, the operating system of the actors and their relationships, and the technical platform).

A fourth extension of the above explanations of joint value creation concerns the consideration of digitalization in joint value creation, which can be improved via the exchange of partners on digital platforms [14, 46] which builds the interface through which actors interact and share resources. The exchange platform can be both, an online and offline infrastructure. Often an app is used, but it is usually only the tip of the mountain of data behind it. According to Hagiu, Rothman [44] there are also six "platform effects" that drive joint value creation (see [5], endnotes 1 and 3):

1. Modularization: Digitalization facilitates joint value creation through modularization [49], i.e. by defining modules and interfaces that reduce transaction costs while creating the flexibility to able to expand the ecosystem [50].
2. Complementarity: Digitalization enables the use of complementarities in production [11], i.e. supply-side economies of scope [42] through the shared performance of tasks or transfer of resources and competencies by the providers.
3. Value co-creation with customers: Digitalization also enables the use of complementarities in consumption [11, 42], i.e. demand-side economies of scope through value co-creation with customers (see endnote 2).
4. Scaling in the network: Digitalization improves joint value creation through network effects (demand-side economies of scale, [51]), e.g. by scaling demand without additional costs.
5. Cross-subsidizing of weak partners: Digitalization improves joint value creation through pricing effects in two-sided markets [14, 52], i.e. by being able to subsidize the more price-sensitive side of the platform.
6. AI-based learning: Digitalization improves the joint value creation through data-based learning, above all through the use of AI [14, 53].

These platform effects increase the joint value creation in ecosystems. As value drivers, they are a fifth key construct of ecosystems. In using these key constructs when designing ecosystems, joint value creation and individual value capture by the partners must be jointly planned and subsequently consolidated [54, 55]: An ecosystem is thus developed in two phases (see [5], endnotes 1 and 3): first a "design phase" which comprises the design of an ecosystem blueprint, then an "alignment phase" in which the blueprint has to be further improved by even stronger networking between the partners (cf. Fig. 2.4). Here, the blueprint provides the framework for the design of the ecosystem, in particular the structure, activities and orchestration and a plan for how value is to be created [56]. It comprises the five previously explained key constructs as process steps starting from specific customer information: first [1] the focal value proposition, [2] the platform and [3] the operating model as the necessary prerequisites for joint value creation and individual value capture in an ecosystem and in addition the [4] governance mechanisms and [5] value drivers.

However, even the design phase is a fairly long process with feedback loops between the process steps, which – at least with capital-intensive partners such as manufacturing firms [57] – often have to be gone through at least twice before a blueprint of an ecosystem is developed [54, 55]. Then it is necessary to establish a minimum viable ecosystem first [9, 55], because given the risks mentioned above [37], ecosystems often do not create value but destroy it (see Fig. 2.3).

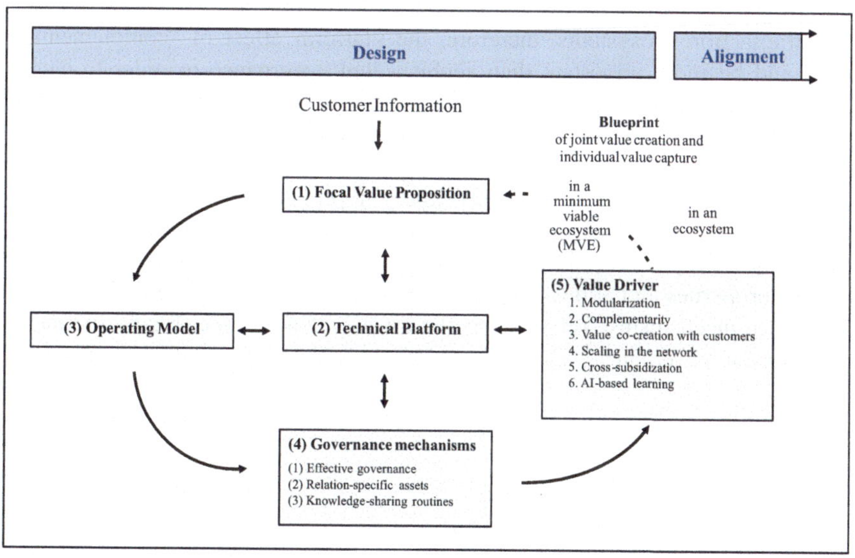

Fig. 2.4 Designing an ecosystem (based on [53] [9], see also endnote 3)

Since this book will examine the transformation towards SDVs and towards shared and connected mobility solutions, a distinction must be made between innovation and transaction ecosystems. Therefore, the last section of this chapter explains that the joint value creation of the partners in these ecosystems differs.

2.4 Different Governance Mechanisms and Value Drivers needed for Joint Value Creation in Innovation and Transaction Ecosystems

A distinction between transactional and innovation ecosystems can be important for joint value creation and individual value capture. Value can be created directly in transaction ecosystems, whereas in innovation ecosystems it is first a matter of jointly developing inventions ("a rising tide lifts all boats", ([58], p. 478), which only become profitable innovations in a second step. Therefore, uncertainty is much higher in innovation ecosystems because the value proposition is not known from ex ante ("de novo ecosystems", [54]) Here it is important to first make a clear potential analysis "narrowing down the future" ([54], p. 477). Consequently, it should be more difficult for the initiator of an innovation ecosystem to win partners because there is no clear vision and no compelling "blueprint" of the future ecosystem that could signal to potential partners that there is a danger of "missing the train" and that they should therefore join the network early ([54], p. 467).

It is assumed here that differences in the factors influencing joint value creation and individual value capture in ecosystems arise depending on the economies of scope (in accordance with [43]) existing between the partners. In the transaction ecosystem, this means joint task performance using shared assets, whereas the innovation ecosystem depends on a transfer of know-how as a quasi-public good.

In transaction ecosystems, therefore, the platform effect of "complementarity" should be more important than sophisticated governance in order to enable exchanges. In addition, rapid cross-subsidization of weak partners is likely to be helpful, because this enables the joint value creation and thus also individual value capture [59].

In innovation ecosystems, it is particularly important to exchange a great deal of know-how even at an early stage [2]. However, there is the risk that knowledge becomes a public good in the ecosystem, causing capabilities to lose their strategic properties and allowing competitors to access them more easily [60]. The governance mechanisms "relation-specific assets", "knowledge-sharing routines " and "effective governance " should therefore be established as quickly as possible.

These theoretical considerations were tested in an empirical investigation in the automotive industry. As the automotive industry is not yet very active in the ecosystem game, these are findings from an early stage in the design of ecosystems. Two research sub-models were identified: one for a transaction ecosystem and one for an innovation ecosystem (for the latter see also endnote 1). Each consisted of several variables that are interconnected by multivariate relationships and could be statistically tested by structural equation analysis [61].

After operationalizing the individual constructs, they were recorded using a seven-point Likert scale ranging from 1 ("I disagree") to 7 ("I strongly agree") [62]. The survey was conducted in the spring of 2022 using a standardized online questionnaire. It was pre-tested with 20 companies to ensure validity [61]. Data from 286 automotive companies were therefore generated (response rate 9.5% of 3000 companies contracted), that describe themselves as active in at least one ecosystem according to Adner's structural definition. The respondents were top managers or people from the strategy departments.

Of the 286 automotive companies, 17% are headquartered in Asia, 33% in North America and the remaining 50% in Europe. Most of the respondents are suppliers (97%) and the remaining 3% OEMs. The turnover is less than US$1 billion for 111 (39%) companies, 103 (36%) have a turnover of between US$1–5 billion and 72 companies (25%) have a turnover of US$5 billion or more. The results are shown in Fig. 2.5.

A multigroup analysis confirms that the effect from individual value capture on rents is significantly higher in the transaction ecosystem (0.32) than in the innovation ecosystem (0.26). This is not surprising because joint value created in a transaction ecosystem is immediately passed on to the rents, whereas in the innovation ecosystem an invention requires a new product, the sale of which then generates rents. The main results are (see endnote 1):

1. Only few value drivers (complementarity and cross-subsidization) and governance mechanisms (effective governance and relation-specific assets) currently have an immediate impact on joint value creation and on individual value capture. Therefore, they seem to be used early in the design of an ecosystem.
2. In both types of ecosystems, joint value creation is significantly influenced by relation-specific assets. Once these relationships are created, value is also created. In contrast, the two types of ecosystems differ in the factors which significantly influence individual value capture. In transaction ecosystems, the partners' share of the jointly created value depends strongly on their complementarity and on cross-subsidization. In innovation ecosystems, relation-specific assets and knowledge transfer routines are crucial.
3. In both ecosystems, joint value creation has no significant effect on value capture. Similarly to Dyer et al. (see endnote 2), no significant sequence of joint value creation and individual value capture is shown. Although a tendency appears, there may actually be a trade-off between maximum value creation and effective value capture [63].

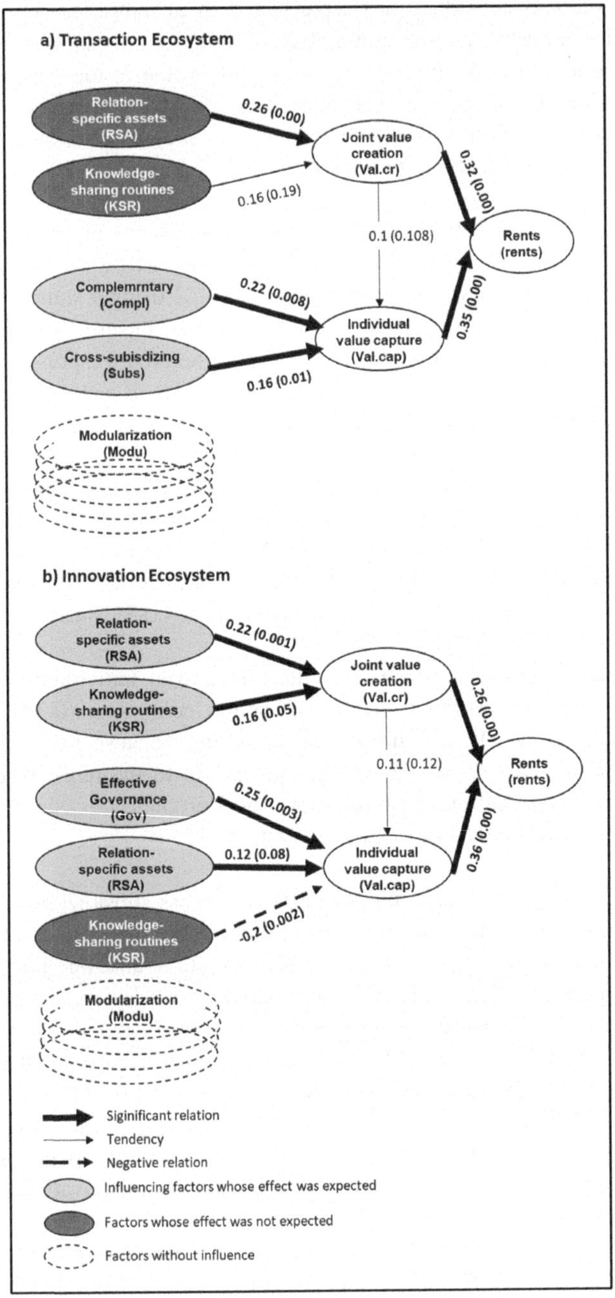

Fig. 2.5 Path coefficients in the research model. *Source* Own design, see also endnote 1

4. The effect of value capture on rents is higher than that of joint value creation, so that the challenge for profitable and thus durable ecosystems in turbulent times seems to lie less in joint value creation and more in individual value capture.

The main implication from this survey for further research is, that the design of ecosystems "evolves" in several loops or phases ([54, 55] and Fig. 2.4), because this static, one-time study showed that the individual value drivers and governance mechanisms have a different influence on joint value creation and individual value capture across all manufacturing companies surveyed. A key implication for business practice can be derived: building trust between partners is important in the development of ecosystems, especially when companies from traditional industries join innovation ecosystems. This is especially true for companies from the automotive industry, which like to dominate and therefore often fail in cooperations. It is important, especially for capital-intensive manufacturing companies, to have a case from the beginning which promises them individual value capture through joint value creation, i.e. a minimum viable solution.

Overall, this chapter has shown that the increasing coopetition between automotive companies for joint value creation takes place in two very different types of ecosystems: In innovation and transaction ecosystems with different overarching value propositions, different actors and a different function of the technical platforms, different governance mechanisms and value drivers.

References

1. Bengtsson, M., & Raza-Ullah, T. (2016). A systematic review of research on coopetition: Toward a multilevel understanding. *Industrial Marketing Management, 57*(3), 23–39.
2. Adner, R., & Feiler, F. (2019). Interdependence, perception, and investment choices: An experimental approach to decision making in innovation ecosystems. *Organization Science, 30*(1), 109–125.
3. Krylova, K. O., Vera, D., & Crossan, M. (2016). Knowledge transfer in knowledge intensive organizations: The crucial role of improvisation in transferring and protecting knowledge. *Journal of Knowledge Management, 20*(5), 1045–1064.
4. Brandenburger, A. M., & Stuart, H. W. (2007). Biform games. *Management Science, 53*(4), 537–549.
5. Kraus, L., & Proff, H. (2024). Tech play and global play in the automotive industry – Mediating goal conflicts with platform-based ecosystems. *International Journal of Automotive Engineering and Management, 24*(1), 1–26.
6. Jeppe, A., Kraus, L., & Proff, H. (2024). Joint value creation and individual value capture in automotive innovation ecosystems. *International Journal of Automotive Technology and Management, 24*(4), 420–440.
7. Leclercq, T., Hammedi, W., & Poncin, I. (2016). Ten years of value cocreation: An integrative review. *Recherche et Applications en Marketing* (English Edition), 31(3), 26–60.
8. Adner, R. (2017). Ecosystem as structure. *Journal of Management, 43*(1), 39–58.
9. Adner, R. (2021). *Winning the right game: How to disrupt, defend, and deliver in a changing world*. MIT Press.
10. Kapoor, R. (2018). Ecosystems: Broadening the locus of value creation. *Journal of Organization Design, 7*(1), 1–16.

11. Jacobides, M. G., Cennamo, C., & Gawer, A. (2018). Towards a theory of ecosystems. *Strategic Management Journal, 39*(8), 2255–2276.
12. Hawley, A. H. (1986). *Human ecology: A theoretical essay*. University of Chicago Press.
13. Moore, J. F. (1993). Predators and prey: A new ecology of competition. *Harvard Business Review, 71*(3), 75–86.
14. Cusumano, M. A., Gawer, A., & Yoffie, D. B. (2019). *The Business of platforms: Strategy in the age of digital competition, innovation, and power*. HarperCollins Publishers.
15. Gawer, A. (2014). Bridging differing perspectives on technological platforms: Toward an integrative framework. *Research Policy, 43*(7), 1239–1249.
16. Giustiziero, G., Kretschmer, T., Somoya, D., & Wu, B. (2022). Hyperspecialization and hyperscaling: A resource-based theory of the digital firm. *Strategic Management Journal, 44*(6), 1391–1424.
17. Nambisan, S., Zahra, S. A., & Luo, Y. (2019). Global platforms and ecosystems: Implications for international business theories. *Journal of International Business Studies, 50*(9), 1446–1486.
18. Ferreira, F. N. H., Proença, J. F., Spencer, R., & Cova, B. (2013). The transition from products to solutions: External business model fit and dynamics. *Industrial Marketing Management, 42*(7), 1093–1101.
19. Ketonen-Oksi, S., & Valkokari, K. (2019). Innovation ecosystems as structures for value co-creation. *Technology Innovation Management Review, 9*(2), 25–35.
20. Sjödin, D., Parida, V., & Visnjic, I. (2021). How large manufacturers digitalize their business models? A framework for orchestrating industrial ecosystems. *California Management Review, 64*(3), 1–29.
21. Adner, R., & Lieberman, M. (2021). Disruption through complements. In: Strategy. *Science, 6*(1), 91–109.
22. Teece, D. J., Peteraf, M., & Leih, S. (2016). Dynamic capabilities and organizational agility: Risk, uncertainty, and strategy in the innovation economy. *California Management Review, 58*(4), 13–35.
23. Jacobides, M. G., & Lianos, I. (2021). Regulating platforms and ecosystems: An introduction. *Industrial and Corporate Change, 30*(5), 1131–1142.
24. Cha, H. (2020). A paradigm shift in the global strategy of MNEs towards business ecosystems: A research agenda for new theory development. *Journal of International Management, 26*(3), Article 100755.
25. Rochet, J.-C., & Tirole, J. (2003). Platform competition in two-sided markets. *Journal of the European Economic Association, 1*(4), 990–1029.
26. Karlsson, I., Mukhtar-Landgren, D., Smith, G., Koglin, T., Kronsell, A., Lund, E., Sarasini, S., & Sochor, J. (2020). Development and implementation of Mobility-as-a-Service – A qualitative study of barriers and enabling factors. *Transportation Research Part A: Policy and Practice, 131*, 283–295.
27. Hietanen, S. (2014). Mobility as a Service – the new transport model? *Eurotransport, 12*(2), 2–4.
28. Kamargianni, M., & Matyas, M. (2017, January 8–12). *The business ecosystem of Mobility-as-a-Service* [Paper presentation]. 96th Transportation Research Board (TRB) Annual Meeting, Washington, DC, United States.
29. Guidon, S., Wicki, M., Bernauer, T., & Axhausen, K. (2020). Transportation service bundling – For whose benefit? Consumer valuation of pure bundling in the passenger transportation market. *Transportation Research Part A: Policy and Practice, 131*(3), 91–106.
30. Alyavina, E., Nikitas, A., & Tchouamou Njoya, E. (2020). Mobility as a service and sustainable travel behaviour: A thematic analysis study. *Transportation Research Part F: Traffic Psychology and Behaviour, 73*, 362–381.
31. Luo, Y. (2021). A general framework of digitalization risks in international business. *Journal of International Business Studies, 53*, 344–361.
32. Van den Berg, V. A., Meurs, H., & Verhoef, E. T. (2022). Business models for Mobility as a Service (MaaS). *Transportation Research Part B: Methodological, 157*, 203–229.

33. Talmar, M., Walrave, B., Podoynitsyna, K. S., Holmström, J., & Romme, A. G. L. (2020). Mapping, analyzing and designing innovation ecosystems: The ecosystem pie model. *Long Range Planning, 53*(4), Article 101850.
34. Brandenburger, A. M., & Stuart, H. W. (1996). Value-based business strategy. *Journal of Economics & Management Strategy, 5*(1), 5–24.
35. Bowman, C., & Ambrosini, V. (2000). Value creation versus value capture: Towards a coherent definition of value in strategy. *British Journal of Management, 11*(1), 1–15.
36. Ozcan, P., & Santos, F. M. (2015). The market that never was: Turf wars and failed alliances in mobile payments. *Strategic Management Journal, 36*(10), 1486–1512.
37. Reeves, M., Lotan, H., Legrand, J., & Jacobides, M. G. (2019). How business ecosystems rise (and often fail). *MIT Sloan Management Review, 60*(4), 1–6.
38. Gans, J., & Ryall, M. D. (2017). Value capture theory: A strategic management review. *Strategic Management Journal, 38*(1), 17–41.
39. Adegbesan, J. A. (2009). On the origins of competitive advantage: Strategic factor markets and heterogeneous resource complementarity. *Academy of Management Review, 34*(3), 463–475.
40. Vargo, S. L., & Lusch, R. F. (2004). Evolving to a new dominant logic for marketing. *Journal of Marketing, 68*(1), 1–17.
41. Ross, D. G. (2018). Using cooperative game theory to contribute to strategy research. *Strategic Management Journal, 39*(11), 2859–2876.
42. Dyer, J. H., Singh, H., & Hesterly, W. S. (2018). The relational view revisited: A dynamic perspective on value creation and value capture. *Strategic Management Journal, 39*(12), 3140–3162.
43. Teece, D. J. (1982). Towards an economic theory of the multiproduct firm. *Journal of Economic Behavior & Organization, 3*(1), 39–63.
44. Hagiu, A., & Rothman, S. (2016). Network effects aren't enough. *Harvard Business Review, 94*(4), 65–71.
45. Weber, C., Bauke, B., & Raibulet, V. (2016). An empirical test of the relational view in the context of corporate venture capital. *Strategic Entrepreneurship Journal, 10*(3), 274–299.
46. Cusumano, M. A., Gawer, A., & Yoffie, D. B. (2021). Can self-regulation save digital platforms. *Industrial and corporate change, 30*(5), 1259–1285.
47. Powell, W. W., Koput, K. W., & Smith-Doerr, L. (1996). Interorganizational collaboration and the locus of innovation: Networks of learning in biotechnology. *Administrative science quarterly, 41,* 116–145.
48. Jaspers, D., Knobbe, F., Proff, H., Schmidt-Szybisty, G., & Salmen, S. (2024). Innovating mobility: Blueprint of a charging ecosystem for electric axis – A mobility study. *International Journal of Strategic Management, 24*(1), 5–25.
49. Baldwin, C. Y., & Clark, K. B. (1997). Managing in an age of modularity. *Harvard Business Review, 75*(5), 84–93.
50. Williamson, P. J., & de Meyer, A. (2012). Ecosystem advantage: How to successfully harness the power of partners. *California Management Review, 55*(1), 24–46.
51. Katz, M. L., & Shapiro, C. (1985). Network externalities, competition, and compatibility. *In. The American Economic Review, 75*(3), 424–440.
52. Eisenmann, T., Parker, G., & Van Alstyne, M. (2006). Strategies for two-sided markets. *Harvard Business Review, 84*(10), 92–101.
53. Iansiti, M., & Lakhani, K. R. (2020). *Competing in an age of AI: Strategy and leadership when algorithms and networks run the world*. Harvard Business Review Press.
54. Dattée, B., Alexy, O., & Autio, E. (2018). Maneuvering in poor visibility: How firms play the ecosystem game when uncertainty is high. *Academy of Management Journal, 61*(2), 466–498.
55. Lewrick, M. (2021). *Business Ökosystem Design.* Vahlen.
56. Spaniol, M. I., & Rowland, N. J. (2022). Business ecosystems and the view from the future: The use of corporate foresight by stakeholders of the Ro-Ro shipping ecosystem in the Baltic Sea Region. *Technological Forecasting & Social Change, 184,* Article 121964.

57. Leppänen, P., George, G., & Alexy, O. (2021). When do novel business models lead to high performance? A configurational approach to value drivers, competitive strategy, and firm environment. *Academy of Management Journal, 66,* 164–194.
58. Khanagha, S., Ansari, S., Paroutis, S., & Oviedo, L. (2022). Mutualism and the dynamics of new platform creation: A study of Cisco and fog computing. *Strategic Management Journal, 43*(3), 476–506.
59. Rietveld, J., Schilling, M. A., & Bellavitis, C. (2019). Platform strategy: Managing ecosystem value through selective promotion of complements. *Organization Science, 30*(6), 1232–1251.
60. Ralston, P., & Blackhurst, J. (2020). Industry 4.0 and resilience in the supply chain: A driver of capability enhancement or capability loss? *International Journal of Production Research, 58*(16), 5006–5019.
61. Hair, J. F., Black, W. C., Babin, B. J., & Anderson, R. E. (2019). *Multivariate data analysis* (8th ed.). Cengage Learning EMEA.
62. Jamieson, S. (2004). Likert scales: How to (ab)use them. *Medical Education, 38*(12), 1217–1218.
63. Schmeiss, J., Hoelzle, K., & Tech, R. P. G. (2019). Designing governance mechanisms in platform ecosystems: Addressing the paradox of openness through blockchain technology. *California Management Review, 62*(1), 121–143.

Technical Architecture of Ecosystems: Customer Application, Middleware and Backend

3

Arman Arzani, Marcus Handte, and Pedro J. Marrón

Contents

Abstract

In an increasingly interconnected world, the concept of digital ecosystems has emerged as a powerful paradigm for delivering seamless, user-centric services across diverse domains. This chapter explores the technical architecture that underpins such ecosystems, focusing on two transformative examples: SDVs and MaaS. These domains exemplify the shift from siloed, monolithic systems to dynamic, data-driven platforms that integrate multiple service providers into a unified experience for end users.

A. Arzani · M. Handte · P. J. Marrón (✉)
University of Duisburg-Essen, Duisburg/Essen, Germany
e-mail: pjmarron@uni-due.de

A. Arzani
e-mail: arman.arzani@uni-due.de

M. Handte
e-mail: marcus.handte@uni-due.de

H. Proff et al. (eds.), *Future of the Automotive Industry*, Future of Business and Finance, https://doi.org/10.1007/978-3-032-26308-7_3

3.1 Introduction

In an increasingly interconnected world, the concept of digital ecosystems has emerged as a powerful paradigm for delivering seamless, user-centric services across diverse domains. This chapter explores the technical architecture that underpins such ecosystems, focusing on two transformative examples: SDVs and MaaS. These domains exemplify the shift from siloed, monolithic systems to dynamic, data-driven platforms that integrate multiple service providers into a unified experience for end users.

At the heart of this transformation lies a high-level architectural framework designed to facilitate interoperability while minimizing interdependencies between ecosystem participants. This architecture enables the aggregation and harmonization of data from disparate sources – ranging from vehicle telemetry and traffic systems to ride – sharing platforms and public transit networks – into a coherent, responsive service layer. By abstracting the complexity of individual systems, the architecture empowers developers and service providers to innovate rapidly without being constrained by rigid integration requirements.

A key enabler of this flexibility is the use of modular data connectors, which serve as interfaces between the core ecosystem and external data sources. These connectors decouple service logic from data ingestion, allowing new providers to be integrated with minimal friction and existing services to evolve independently. This approach not only enhances scalability and resilience but also fosters a vibrant ecosystem where innovation can thrive. By presenting a detailed example from the MaaS domain, this chapter illustrates how such an architecture can support the creation of intelligent, adaptive mobility solutions that are greater than the sum of their parts.

The remainder of this chapter is structured as follows. In Sect. 3.2, we first review some existing MaaS and SDV architectures that have been presented in literature. Based on this, we outline a prototypical architecture for integrating the data from different service providers into a unified user experience in Sect. 3.3. To demonstrate the practical application of this architecture, we describe its use to create a unified MaaS experience in the InnaMoRuhr research project in Sect. 3.4. Finally, Sect. 3.5 concludes the chapter with a summary.

3.2 Related Work

Various practitioners and researchers have proposed system architectures for digital ecosystems that target MaaS and SDVs. These system architecture define the structure of an ecosystem by specifying its core components and assigning responsibilities or requirements to them. In this section, we examine several existing architectural models from the literature, focusing on how they address the integration in MaaS and SDV.

Several architectures have been proposed to support MaaS ecosystems, each emphasizing different technical and organizational priorities. The Siemens architecture [18] introduces a B2B platform that streamlines integration between transport services and frontends, enabling unified access to routing, billing, and ticketing through open APIs. Similarly, the SMILE project [10] tested a smartphone-based mobility app, emphasizing seamless service access and partner integration via standardized connectors. The Datson reference model [3] extends this by incorporating broader domains such as consumer needs, business logic, technology, and regulatory policies, while clarifying the roles of core stakeholders. The MAASiFiE architecture [12] adds a four-level model, highlighting governance and regulatory dimensions crucial for scalable MaaS deployment in Europe. Other projects such as HoPE [6] focus on eco-mobility, offering users multimodal planning, secure payments, and environmentally conscious routing in various European cities. Expanding the architectural landscape, Ambrosino et al. [1] proposed a FSUM Centre model to coordinate public and private mobility services, enabling integrated trip planning, booking, and payment across collective and individual transport modes. The Mobility Broker [2] was an early concept that unified service data via standardized interfaces. Pflügler et al. [16] offered a modular, layered approach aimed at enabling smart city solutions through open data use. Moving beyond traditional MaaS, the eMaaS architectures [15, 17] extend the ecosystem to include electric mobility services such as e-scooters, e-bikes, and EV fleets. These systems introduce components for smart charging, fleet optimization, and vehicle-to-grid (V2G) integration, addressing both sustainability and the specific operational needs of electric mobility.

Furthermore, SDVs signify a paradigm shift in vehicle architecture by decoupling hardware from software, thereby enabling continuous updates, personalization, and service evolution through OTA mechanisms [4, 13]). Architecturally, SDVs transition from distributed networks of Electronic Control Units (ECU) toward centralized and zone-based models, with high-performance computing platforms, often System on Chips (SoC), handling a broad range of tasks. These platforms support a layered software architecture comprising a foundational OS kernel, middleware for inter-component communication, and a service-oriented architecture (SOA) layer that modularizes stable vehicle functions as reusable services. On top lies application software, including both functional (e.g. control, safety) and service-oriented (e.g. infotainment, energy) modules. These onboard systems interface with cloud platforms to enable a feedback loop for data processing, algorithm training, and remote maintenance. Extending these principles, the Software-Defined Internet of Vehicles (SD-IoV) introduces an SDN-based architecture with logically centralized or hierarchically distributed controllers, SDN-enabled vehicles, wireless and wired access networks, and a layered control and data plane system [5, 11, 19]. This architecture emphasizes dynamic, centralized control over heterogeneous vehicular networks and supports real-time responsiveness via a dedicated knowledge plane. Historically fragmented software stacks in vehicles underscore the growing need for integrated, end-to-end software platforms that can unify development environments, simplify interfaces, and enhance system security and performance.

Although, MaaS and SDV are addressing different domains, many of the existing ecosystem architectures share common elements, such as the focus on data sharing and infrastructure, platform functionalities, and user-centric design. In all architectures, data is gathered from different service providers and integrated into the platform. Once this data has been processed and integrated, the mobility services can be offered as a unified experience to end users and other stakeholders such as service providers, and platform operators.

3.3 Architecture

Based on the main functions that are typically encompassed in existing digital ecosystem architectures in the MaaS and SDV domain as well as our own experience with integrating the mobility offers from different transportation service providers in several projects [7, 9, 20], in this section, we discuss a minimal yet extensible ecosystem architecture. To do this, we first present an overview of the architecture and decompose it into individual layers. Thereafter, we describe the functions and components of each layer. Finally, we discuss the interaction and communication among them.

3.3.1 Layers

Conceptually, many business applications can be decomposed into a three-layer architecture. As indicated by its name, this architecture separates an application into three distinct layers, each with a specific responsibility. The presentation layer at the top is responsible for the user interface and interaction, which could be a web page, mobile app, desktop or a vehicle dashboard interface where users input data and receive output. The application layer in the middle handles the business rules and logic of the application. It processes user inputs, applies the necessary logic, and determines how data should flow between the presentation and data layer. The data layer at the bottom is responsible for data storage and management. It interacts with a database or other data storages and manages the retrieval and storage of information.

For digital ecosystems, however, we need to consider the fact that service providers are often already operating their own systems in order to manage their data and offer their service. To account for this, we introduce an additional intermediary layer that connects the ecosystem related business logic with the systems of service providers through a connector. Since the operation of the platform itself may also generate data, we also add a data storage that can be accessed directly. Figure 3.1 shows the resulting architecture, including the mapping to a traditional three-layer architecture as well as the interaction between the different layers. In the following, we describe each layer in more detail.

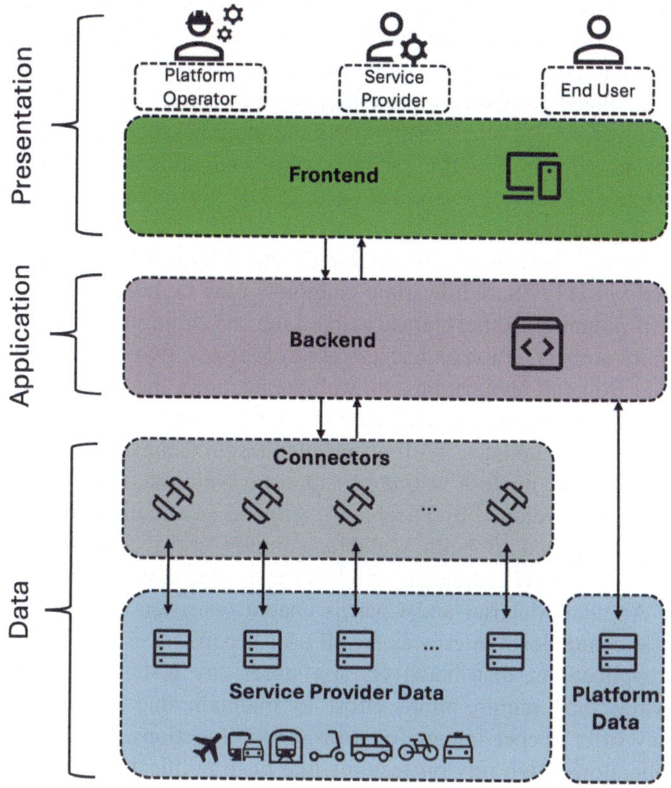

Fig. 3.1 Minimal and extensible architecture

Frontend

The frontend layer serves as the interface through which users interact with underlying ecosystem platform. It enables access to key functionalities such as trip planning, vehicle control, and service customization, while delivering real-time updates such as vehicle availability, route changes, or system notifications. In ecosystem architectures today, the frontend is usually implemented using mobile platforms, web-based technologies, or embedded in-vehicle interfaces such as digital dashboards, infotainment systems, or head-up displays, particularly in the case of SDV.

Since ecosystems involve different user groups, the frontend layer may entail a number of separate applications, each targeted to a specific user role. Existing systems usually address at least *platform operators*, *service providers*, and *end users*. End users are individual travelers or businesses who interact directly with the mobility services by planning trips, accessing vehicles, or managing subscriptions, typically through mobile apps or in-vehicle interfaces. These applications allow users to manage account information, plan and book trips, visualize route options, and receive contextual updates such as disruptions or alternative route proposals.

Platform operators, such as car manufacturers, integrated mobility providers or, city transport authorities, are responsible for orchestrating the ecosystem, integrating services, managing user data, and maintaining system-wide coordination. They often use specialized dashboards to monitor system metrics and analytics outputs. Service providers, on the other hand, contribute modular services that are embedded into the broader ecosystem, ranging from navigation (e.g. Google Maps), infotainment (e.g. Spotify), and ticketing (e.g. DB Navigator), to data analytics and cloud infrastructure.

To access backend services, frontend applications typically rely on web-based APIs secured via HTTPS. While some endpoints may be publicly accessible (for example, trip planning), others, such as booking, reservations, or operator dashboards, require authentication and access control. Today, two common approaches to developing frontend applications are web applications and mobile applications.

Web applications are executed inside a web browser and do not require installation, allowing for fast updates. With responsive design techniques and advances in browser performance, modern web apps support a wide range of devices. Frameworks like React (developed by Facebook), Angular (maintained by Google), and Vue.js are widely used to build scalable, component-driven UIs. Additionally, libraries such as DeckGL, Leaflet, or OpenLayers support geospatial visualization, while Angular Material and Chart.js enable consistent design systems and data-rich dashboards level interfaces for all user groups.

Mobile applications, distributed via app stores, are installed directly on user devices. While they require more effort to maintain due to platform-specific updates, they offer deeper integration with device functions. For MaaS systems, mobile applications often rely on features like push notifications (via Firebase or APNS) to deliver timely updates, secure persistent storage to manage offline credentials and tickets, and access to motion/location sensors to enable automated trip detection or usage-based billing. Furthermore, mobile app development is typically supported by native (Software Development Kits) SDK provided by platform owners (e.g. Apple or Google), but this introduces challenges when developing for multiple platforms. To reduce development overhead and increase code reuse, cross-platform frameworks like Xamarin, Flutter, or React Native are used.

In the context of SDVs, the frontend inside the vehicle is primarily realized through software interfaces such as digital instrument clusters, infotainment systems, and head-up displays, which are often developed using embedded operating systems (e.g. QNX, Android Automotive) and automotive-grade UI frameworks (e.g. Qt, Kanzi) [14]. These interfaces provide users with direct access to vehicle functions, such as climate control, navigation, media, and driver-assistance features. In addition, mobile applications extend the vehicle interface by enabling remote access to vehicle status, route pre-conditioning, or charging management.

Backend

The backend layer in digital ecosystems serves as the core infrastructure that implements platform functionality and provides interfaces to the frontend layer. It acts as the central hub that processes API requests, performs logic execution, and accesses necessary resources to deliver services such as trip planning, vehicle control, and data synchronization. For mobile and web-based applications, these functionalities are typically exposed via RESTful API endpoints, enabling interoperability across frontend technologies. In SDV architectures, similar service endpoints may be used to interface with in-vehicle systems, allowing seamless data flow between backend intelligence and embedded systems.

To interact with external mobility and transport service providers, the backend uses dedicated connectors. The backend is also tasked with maintaining overall system efficiency, security, and reliability. This includes safeguarding data through encryption, implementing access control mechanisms, and ensuring compliance with regulatory frameworks such as the General Data Protection Regulation (GDPR).

A key component of backend security is authentication, which verifies the identity of users, applications, or services before granting access to protected resources. Typically, this involves a trusted authentication server that validates credentials and issues cryptographically signed tokens, such as JSON Web Tokens (JWTs). These tokens are then used by the resource servers to determine access rights without repeatedly prompting the user for login. JWTs may contain role-specific claims that allow fine-grained access control, for instance, distinguishing between end users, platform operators, and service providers. By separating the authentication server from the resource server, the architecture enhances scalability and enables advanced security features like multifactor authentication or single sign-on, without burdening service logic with credential management.

Beyond core functionality, the backend also usually includes a powerful analytics component, which plays a role in monitoring platform health, improving service quality, and enabling intelligent personalization. This component collects and processes vast amounts of structured and unstructured data, such as trip histories, vehicle telemetry, user preferences, and external conditions (e.g. weather or traffic). Tools like Apache Hadoop and Apache Spark are often employed to manage large-scale data processing. While Hadoop offers scalable batch processing via its MapReduce framework and distributed file system (HDFS), Spark supports real-time and iterative computations in memory, suitable for applications like predictive maintenance, dynamic fleet allocation, or anomaly detection in vehicle operations.

AI-driven modules built on top of these frameworks can further enhance platform intelligence. For example, demand prediction models can forecast peak usage times and optimize resource allocation accordingly. Route optimization algorithms can dynamically reassign vehicles based on live traffic data, while personalization engines tailor recommendations to individual users, improving engagement and retention. These models are especially critical in SDV ecosystems, where adaptive behavior, such as context-aware navigation, autonomous vehicle dispatching, or user-specific infotainment experiences, can enhance user satisfaction and operational efficiency.

To support these functions at scale, modern backends are often implemented as distributed systems using microservices architectures and cloud-native infrastructure. The modular development ensures resilience against failures, and elastic scalability in response to fluctuating demand, which is essential for both urban-scale MaaS operations and large SDV fleets.

Connectors
The connectors take care of the required provider-specific data integration. Besides from implementing the data access mechanisms and format conversions, the connectors may also validate the plausibility of incoming data. Thereby, the validation may range from simple format to application-specific consistency checks. An example for the latter is validating that the location of a shared vehicle actually lies inside the operator's service zone or that a reported sensor value of a vehicle's engine lies within a feasible range. Depending on the data type and the required update rate, connectors can often apply some form of caching in order to ensure that the platform remains operational, if a provider system becomes temporarily unavailable.

3.3.2 Interaction

Communication between the different layers of a mobility ecosystem architecture, such as those involving MaaS platforms and SDVs, is typically facilitated through communication middleware that supports both request-response (pull) and publish-subscribe (push) interaction patterns. A widely adopted approach for pull-based communication is to expose the APIs of various components via RESTful services. In this model, interaction between layers is realized via HTTP requests (e.g. GET, POST, PUT, DELETE), allowing clients (mobile apps, web portals, or in-vehicle systems) to fetch, create, update, or delete data on demand. For example, a MaaS application may issue a GET request to retrieve a trip plan, prompting the backend to query real-time data from multiple mobility providers (e.g. transit schedules, vehicle availability), compute optimal routes, and return the result in a structured format like JSON. Similarly, SDV systems may request backend services to retrieve map tiles, receive navigation updates, or synchronize user profiles. This pull-based mechanism is ideal for services that are user-initiated or context-dependent, such as trip planning, booking, payment, or requesting software updates.

Conversely, push-based, message-oriented middleware plays a crucial role in enabling real-time updates across MaaS and SDV components. Technologies such as WebSockets, Apache Kafka, or MQTT are commonly used to implement this pattern. In a MaaS context, push-based messaging supports real-time communication of vehicle delays, route changes, or last-minute cancellations to subscribed user interfaces. In SDVs, this mechanism is vital for streaming high-frequency telemetry data (e.g. vehicle position, sensor updates, or driver alerts) from the vehicle to the backend or to other ecosystem actors like fleet operators. To participate in this communication, consuming components (e.g. an operator dashboard, mobile frontend, or SDV controller) subscribe to specific topics or endpoints, after which relevant messages are automatically pushed to them as they become available. This avoids the need for continuous polling and enables the rapid delivery of time-critical information, such as notifying users about vehicle delays or transmitting emergency alerts to autonomous vehicles.

3.4 Example: InnaMoRuhr

Funded by the Ministry of Transport of Nordrhein-Westfalen, the goal of the InnaMoRuhr research project was to develop a mobility concept for the University Alliance Ruhr, which encompasses 3 universities with approximately 110,000 students and 14,000 scientists located in the Ruhr area. The InnaMoRuhr project ran from 2020 to 2023 and had an interdisciplinary setup involving social scientists, economists, electrical engineers and computer scientists. In addition to studies and large-scale questionnaires, the project also included field trials to study behavioral changes induced by novel mobility options. The options included a monthly mobility budget and free access to a small fleet of electrical cars.

To support the field trials, the project developed a MaaS platform in order to integrate the additional mobility options with the existing offers available in the Ruhr area. This platform enabled the participants of the field trials to plan intermodal trips and to book the electric vehicles using a web application and mobile applications for iOS and Android. To capture the mobility behavior of the participants, the mobile application allowed users to automatically capture their trips and to share the captured data with the scientists.

The resulting platform represents a complete implementation of the digital ecosystems architecture introduced in the previous section. It covers all layers and provides a unified experience to end users via mobile and web applications. In the following, we describe the components and their functions in more detail.

3.4.1 InnaMoRuhr Connectors

The basis for the InnaMoRuhr platform are data connectors that integrate with different mobility service providers. Each data connector is implemented as a RESTful service that can be queried by the backend. Each connector maintains its own data repository that caches the most recent data. In addition, some connectors also store historic data, which is used to generate forecasts. In total, the platform encompasses the following four connectors:

- *Public Transportation*: This connector integrates with the systems of the regional transport association in the Ruhr area (Verkehrsverbund Rhein-Ruhr). The transport association regularly publishes transit schedules in NeTEX and

GTFS format via its open data website. In addition, it operates web-based services to access information on route incidents and delays. Since the schedule data typically remains stable for several weeks, the connector is configured to check for schedule updates once per day. If a new schedule is available, the GTFS schedule is downloaded into the data repository and validated. If the validation completes successfully, the connector triggers an update to subscribed backend components, which can then process and integrate the new data. For incidents and delays, the connector applies a reactive caching strategy that is initiated by route computations inside the planning component. Since the routing algorithm may issue multiple queries for delays during the route computation, the route incidents and delays are cached locally by the connector as well. This reduces the number of queries issued to the external system, and it reduces the latency of route computations.

- *Rental Bikes*: This connector integrates with the main provider for rental bikes that operates numerous bike sharing stations in the Ruhr area. The connector periodically downloads the location of rental stations, as well as the positions and availability of rental bikes in the target area. Due to API restrictions defined by the provider, the requests are limited to 2 updates per hour. In addition to providing access to the latest information, the connector also stores the historic information. Using the resulting history, the connector can generate statistics on bike availability for each hour on each weekday for each station. The expected availability information can then be used during route computation, and it can be displayed to travelers.
- *Rental E-Scooters*: This connector integrates with e-scooter providers offering APIs that follow the GBFS standard. In contrast to rental bikes, which must be returned at stations, e-scooters can typically be rented and returned at any place that lies outside of no parking zones. To handle this, GBFS defines endpoints to access the locations of (available) e-scooters, as well as so-called geo-fencing zones to define restrictions. In addition, GBFS also models rental fees, which can be used to estimate trip costs. For the duration of the field trials in InnaMoRuhr, we were able to attach the connector to one of the major e-scooter providers in the Ruhr area. However, due to legal reasons, we were not allowed to collect historical data. Thus, the connector was configured to discard historical data and was only used as a local cache that was updated once per minute.
- *Electric Cars*: This connector integrates with the fleet management service developed by one of the project partners of InnaMoRuhr who was responsible for maintaining the electric cars. The fleet management system provides access to the real-time position of the vehicles as well as their charging state. The latter includes an estimate on the remaining range. In addition, the fleet management service allows the booking of vehicles. Due to legal constraints, the participants that were interested in using the electric vehicles had to register in advance, and they had to enter contractual agreements (e.g. agree to certain insurance policies). As a result, the booking functions were only available to

registered users and the functions for access control were integrated into the connector.

3.4.2 InnaMoRuhr Backend

On top of the connector layer, the InnaMoRuhr backend integrates the data of the mobility providers and offers a uniform interface for applications. From a travelers' perspective, the main functions of the backend are trip planning and booking. In addition, the backend provides the ability to monitor the behavior of the participants via the built-in analytics. To secure the access to restricted function and sensitive data (i.e. booking and analytics) the backend encompasses an authentication system. Just like the connectors, each component is implemented as a service that exposes its functionality via a REST API. In order to facilitate the implementation of different applications, the APIs are formally specified using an OpenAPI Specification. This enables the automatic generation of API clients for different programming languages and frameworks, which reduces the implementation effort for applications. In the following, we briefly describe the implementation of some of the components.

- *Planning*: The planning component enables travelers to compute and compare intermodal trips from an origin to a destination at a particular point in time that adheres to their mobility preferences. To specify the origin and destination, travelers are typically relying on street addresses. To define preferences, they can customize parameters such as "avoid buses" or "include e-scooters." To compare alternatives, travelers are used to viewing the alternatives in a list representation that highlights specific characteristics (e.g. trip duration, start time, number of transfers) and may also want to display a spatial representation of the trip on a map. Intuitively, this requires the integration of several functions including geocoding (to resolve addresses into coordinates), integrated routing (to compute door to door trips) and map generation (to support the visualization of trips).
- To implement these functions, the InnaMoRuhr platform relies on LocosLab Carta which is a suite of spatial services that leverages open data provided by the OpenStreetMap project. LocosLab Carta ships with algorithms that enable street routing using different profiles (e.g. walking, driving, cycling) and transit routing using GTFS schedules. In addition, it provides a map rendering system with customizable styles and a geocoder that supports autocompletions. InnaMoRuhr uses most of these services but extends the routing service with additional algorithms and models to integrate the sharing providers [9], including rental bikes and e-scooters as well as electric cars.
- To generate route proposals, the planning component integrates with all connectors described previously. Triggered by schedule updates of the public transport connector, the planning component can automatically rebuild its internal models to accommodate changes. When updating the schedule, it integrates other

infrequently changing information such as the position of rental stations for bikes and electric cars as well as zone information for e-scooters. Highly dynamic information such as transit delays, vehicle availability, and predictions are requested from each of the connectors as needed and considered during the route computation.

- *Authentication*: In the InnaMoRuhr platform, trip planning is implemented as a public function that is available to any user. The booking and analytics components described in the following, however, are not publicly accessible, as this could have a negative impact on the service availability, data integrity, as well as the privacy of travelers. The authentication component ensures the requests to the booking and analytics component can only be executed by authorized callers.
- To do this, the component issues JSON Web Tokens that include the caller identity as well as the user's role. New tokens are only issued after a successful credential verification step. Since the tokens are digitally signed, the information contained in them cannot be altered and as a result, the access control logic of the booking and analytics component can trust them, provided that the token has not expired, and the digital signature is valid.
- For access control, the platform differentiates three roles, namely anonymous traveler, registered traveler, and researcher. Anonymous travelers may share their trip information that can be automatically recorded via the mobile application with the analytics component. Registered travelers are able to reserve electric cars required for their trips. In addition, they can cancel reservations that they made earlier, and they can access the description, license plate number, and real-time location of a car during an active reservation period. Researchers can use the analytics component to get aggregated information on the trips of travelers.
- *Booking*: The booking component enables travelers to manage reservations of electric cars for their trips. To do this, it interacts with the fleet management system that realizes the physical access control to the car via the associated connector. Using the connector, the booking component can receive a list of available cars including a description of the brand name, color, and license plate number as well as their real-time location and charging state. The location and charging state are dynamically generated by an embedded system that is installed in each car. To access the car's charging state, the system taps into the internal data via the onboard diagnostics interface. To transmit the data, the system is equipped with network access through a GSM module. Together, the information enables travelers to find and identify the correct car on the road. In addition, it can be used as a selection criterion during the booking process. The management system can also restrict the possible routes that can be taken with a particular car which can be used, e.g. to balance the car distribution in a target area. However, in the context of the field trials of the InnaMoRuhr project, this function was only used to assign a few cars to specific routes that were frequently requested by the participants.

- *Analytics*: Due to the scientific nature of the InnaMoRuhr project, the primary goal of the platform's analytics component was the generation of data to capture the behavior of travelers. To do this, the analytics component is designed to process tracking information that is generated by the mobile application. This tracking information consists of sequences of GPS readings that are collected during trips. To comply with the regulations of the GDPR, the readings are collected only with the explicit consent of the travelers and the origin and destination of the trips are obfuscated on the mobile device. To determine the movement modality of the travelers, the analytics component processes the GPS readings with a hierarchical classifier that has been generated using machine learning. The classifier differentiates several modalities including bus, train, bicycle, e-scooter, car, and walking with a high accuracy, which allows researchers to determine the effects of their experiments. More information on technical approach as well as some scientific results can be found in [8].

3.4.3 InnaMoRuhr Applications

The InnaMoRuhr applications expose the planing and booking functions of the MaaS platform to travelers. To support different usage scenarios, the InnaMoRuhr platform encompasses a web application and a set of associated mobile applications for iOS and Android. The web application is implemented as a single page application using the Angular framework. The mobile applications are written in Java and Swift using the software development kits of the operating system manufacturer. However, to reduce the maintenance effort, the mobile applications embed the web application inside a web container. As a result, changes to the web application are automatically available to the mobile applications without requiring software updates.

To minimize the learning effort, the basic design of the web application follows the patterns used by many popular mapping applications. It displays a large map on the right with an option pane on the left. Depending on the usage, the option pane may show detailed information or a set of controls to search for places, to compute routes, or to book an electric vehicle. To support different screen sizes, the layout adapts dynamically to the available space. On devices with a narrow screen, the option pane and the map are stacked on top of each other, as depicted in Fig. 3.2.

Similar to the layout, the basic workflow for trip planing also follows popular mapping applications. Meaning that a traveler would typically start the trip planning by searching a destination. Once the destination is selected, the app shows additional information about the place which may include, for example, a weather forecast for the next few days as shown in Fig. 3.2a. By tapping a navigation button, the traveler may either compute a route from its current location or specify an alternative origin. Thereafter, the traveler can customize the route computation, e.g. by selecting modalities that should be considered or by specifying a desired departure or arrival time. The resulting routes are displayed in a list view that shows the most important information such as the trip duration, the number of segments with the associated modalities as well as an estimate on the greenhouse gases generated by the trip. By selecting a trip, the traveler can switch to a detailed view that shows additional information, such as the transport provider or information about incidents (c.f. Figure 3.2b). If a trip requires the use of an electric car, the traveler can optionally start the booking process. For rental bikes, the traveler can optionally display additional information on the pickup location, including the predicted availability, as shown in Fig. 3.2c.

As discussed previously, the mobile applications reuse the web application for planning and booking by embedding them in a web container. In addition, they enable travelers to automatically track their trips. In the context of the InnaMoRuhr project, this functionality is primarily used to track the mobility behavior of the participants of the field trials. In other scenarios, however, this function might be useful for demand estimation, personalization, or automatic billing.

(a) Location-based Search and Map with Options

(b) Combining Bike-Sharing and Public Transport

(c) Prediction of Bike Availability for Station

Fig. 3.2 InnaMoRuhr web application for trip planning and booking

In the InnaMoRuhr apps (c.f. Fig. 3.3), the trip tracking is implemented as an optional feature that can be enabled and disabled at any point in time. To meet the guidelines of the app stores, the controls are prominently placed on the main menu which also includes a description of the potential impact on the user's privacy as shown in Fig. 3.3a. To minimize this impact, the mobile applications perform on device trip detection and obfuscation before uploading data to the analytics component. The trip detection segments the GPS readings into individual sequences representing the trips. Once a trip is detected, the obfuscation process removes the exact origin and destination by mapping the nearby readings to grid cells spanning several hundred meters. To enable users to make use of their own tracking data, the application includes statistics as well as visualizations (c.f. Fig. 3.3b) and enables the export of the underlying data. As an additional incentive scheme, the apps also use the tracked data to issue various awards, as shown in Fig. 3.3c. For InnaMoRuhr, this incentive scheme turned out to be a cost-efficient and a highly effective nudge, since many users tried to acquire all available awards resulting in a high participation over the full duration of the trials.

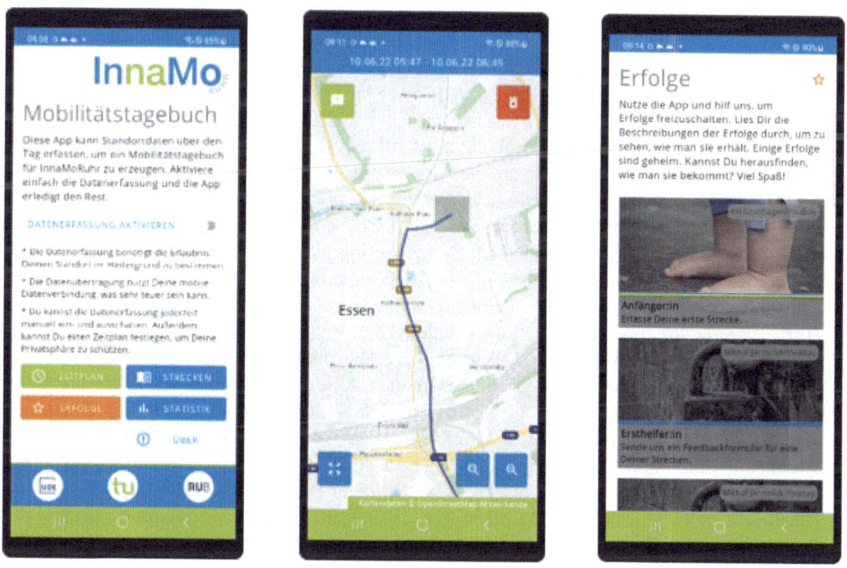

(a) Main Menu with Controls for Automatic Trip Tracking

(b) Visualization of a Trip with Obfuscated Destination

(c) Awards for Trip Tracking to Motivate Participants

Fig. 3.3 InnaMoRuhr mobile application with automatic trip tracking

3.5 Conclusion

In an increasingly interconnected world, the concept of digital ecosystems has emerged as a powerful paradigm for delivering seamless, user-centric services across diverse domains. In this chapter, we have outlined a flexible and scalable architecture for digital ecosystems, with a focus on enabling seamless integration of diverse service providers, which is a fundamental requirement in the domains of SDV and MaaS.

To address the challenge of integrating heterogeneous data sources from multiple services providers into a seamless, user-centric experience, the architecture leverages modular data connectors. The data connectors minimize direct dependencies between components and foster interoperability, rapid innovation, and independent evolution of services. The practical implementation within the InnaMoRuhr research project demonstrates the architecture's effectiveness in creating a unified, user-centric MaaS experience.

References

1. Ambrosino, G., Nelson, J. D., Boero, M., & Pettinelli, I. (2016). Enabling Intermodal Urban Transport Through Complementary Services: From Flexible Mobility Services to the Shared Use Mobility Agency: Workshop 4. Developing Inter-Modal Transport Systems. *Research in Transportation Economics, 59,* 179–184.
2. Beutel, M. C., Gökay, S., Kluth, W., Krempels, K.-H., Samsel, C., & Terwelp, C. (2014). Product oriented integration of heterogeneous mobility services. In *17th International IEEE Conference on Intelligent Transportation Systems* (ITSC) (pp. 1529–1534). IEEE.
3. Datson, J. (2016). *Mobility as a service: Exploring the opportunity for mobility as a service in the UK.*
4. El-Fatyany, A., Wang, X., Duggirala, P. S., Chakraborty, S., Pasricha, S., & Singh, A. K. (2024). Special session: Emerging architecture design, control, and security challenges in software defined vehicles. In *2024 International Conference on Hardware/Software Codesign and System Synthesis (CODES+ ISSS)* (pp. 27–36). IEEE.
5. Fletcher, R., Mahindroo, A., Santhanam, N., & Tschiesner, A. (2020). The case for an end-to-end automotive-software platform. *McKinsey & Company,* 1–16.
6. García Hernández, J. (Ed.). (2014). *Holistic personal public eco-mobility—Deliverable 2.3—HoPE architecture analysis.* https://www.tmleuven.be/en/project/hope
7. Handte, M., Foell, S., Wagner, S., Kortuem, G., & Marrón, P. J. (2016). An Internet-of-Things Enabled Connected Navigation System for Urban Bus Riders. *IEEE Internet of Things Journal, 3*(5), 735–744. https://doi.org/10.1109/JIOT.2016.2554146
8. Handte, M., Kraus, L., & Marrón, P. J., & Proff, H. (2024). Analyzing the mobility of university members for InnaMoRuhr. In H. Proff (Ed.), *Next chapter in mobility: Technische und betriebswirtschaftliche aspekte* (pp. 461–474). Springer Fachmedien Wiesbaden.
9. Handte, M., Kraus, L., Zella, M., Marrón, P. J., & Proff, H. (2023). Integrated Bicycle and Public Transit Routing for InnaMoRuhr. In H. Proff (Ed.), *Towards the New Normal in Mobility: Technische Und Betriebswirtschaftliche Aspekte* (pp. 529–545). Springer Fachmedien Wiesbaden.
10. Heyn, Markus. (2018). Preface: Beyond the hood: The development of mobility services in the mobile internet. *Digital Marketplaces Unleashed,* 457–465.

11. Jiacheng, C., Haibo, Z. H. O. U., Ning, Z., Peng, Y., Lin, G., & Sherman, S. X. (2016). Software Defined Internet of Vehicles: Architecture, Challenges and Solutions. *Journal of Communications and Information Networks, 1*(1), 14–26.
12. König, D., Piri, E., Karlsson, M., Sochor, J., & Heino, I. (2017). *Deliverable 5: Technology for MaaS. MAASiFiE project funded by CEDR.*
13. Liu, Z., Zhang, W., & Zhao, F. (2022). Impact, Challenges and Prospect of Software-Defined Vehicles. *Automotive Innovation, 5*(2), 180–194.
14. Lorenz, M., Amorim, T., Dey, D., Sadeghi, M., & Ebel, P. (2024). Computational models for in-vehicle user interface design: A systematic literature review. In *Proceedings of the 16th International Conference on Automotive User Interfaces and Interactive Vehicular Applications* (pp. 204–215).
15. Nikolaidou, A., Papadopoulos, E., Politis, I., & Basbas, S. (2023). An Indicator-Based Methodological Framework for Assessing an eMaaS Scheme. *World Electric Vehicle Journal, 14*(7), 186.
16. Pflügler, C., Schreieck, M., Hernandez, G., Wiesche, M., & Krcmar, H. (2016). A Concept for the Architecture of an Open Platform for Modular Mobility Services in the Smart City. *Transportation Research Procedia, 19,* 199–206.
17. Reyes García, J. R., Lenz, G., Haveman, S. P., & Bonnema, G. M. (2019). State of the art of Mobility as a Service (MaaS) ecosystems and architectures—an overview of, and a definition, ecosystem and system architecture for Electric Mobility as a Service (eMaaS). *World Electric Vehicle Journal, 11*(1), 7.
18. Siemens. (2016). *SIEMENS MaaS operation and integration with demand-responsive transport in Tampere.* https://tampere.cloudnc.fi/download/noname/%7Bda0fce2f-9084-481c-bf80-94a2f615c86e%7D/1890722
19. Szalavetz, A. (2024). In-House Software Development for Software-Defined Vehicles: Major Changes Ahead in Automotive Value Chains? *International Journal of Automotive Technology and Management, 24*(5), 22–39.
20. Wagner, S., Fet, N., Handte, M., & Marrón, P. J. (2017). An approach for hybrid indoor/outdoor navigation. In *2017 International Conference on Intelligent Environments (IE)* (pp. 36–43). https://doi.org/10.1109/IE.2017.22.

Software-Defined Vehicle Innovation Ecosystems – An Overview

4

Elmar Pritsch, Erik Thorwirth, and Corinna Cruceru-Weisbrod

Contents

Abstract

SDVs form a market that is estimated to be worth US$600 billion in 2030. It is expected that 95 percent of all new vehicles sold by then will be equipped with SDV systems and that automotive software will generate sales of US$84 billion. These estimates show how important the SDV transformation is for the automotive industry. However, it cannot succeed on its own, which is why connected and highly automated Software-Defined Vehicle (SDV) innovation ecosystems are needed.

E. Pritsch (✉) · E. Thorwirth
Deloitte Consulting Germany, Düsseldorf, Germany
e-mail: epritsch@deloitte.de

E. Thorwirth
e-mail: ethorwirth@deloitte.de

C. Cruceru-Weisbrod
Deloitte Consulting Germany, München, Germany
e-mail: CCruceru@deloitte.de

H. Proff et al. (eds.), *Future of the Automotive Industry*, Future of Business and Finance, https://doi.org/10.1007/978-3-032-26308-7_4

4.1 Introduction

SDVs form a market that is estimated to be worth US$600 billion in 2030. It is expected that 95 percent of all new vehicles sold by then will be equipped with SDV systems and that automotive software will generate sales of US$84 billion [1]. These estimates show how important the SDV transformation is for the automotive industry. However, it cannot succeed on its own, which is why connected and highly automated SDV ecosystems are needed.0

In this chapter, we first characterize the design of SDV ecosystems as innovation ecosystems in the automotive industry (Sect. 4.2). We then shed light on the necessary requirements for the vehicle (software and hardware) architecture – onboard (Sect. 4.3) and offboard (Sect. 4.4). Section 4.5 finally shows the importance of scaling in SDV ecosystems, before the chapter ends with looking at the benefits of SDV ecosystems as an outlook in Sect. 4.6.

4.2 Designing SDV Ecosystems as Innovation Ecosystems in the Automotive Industry

While the further spread of shared driving (MaaS, see Chaps. 6 and 7 in this book) and EVs hardly require any further radical technical innovations, but rather clever, innovative business models, often in transaction ecosystems, the transformation to connected and autonomous driving still requires a great many innovations that combine the use of hardware and software on the way to a SDV. Building future-proof SDV ecosystems as innovation ecosystems in the automotive industry require a comprehensive understanding of the key elements shaping the market. They include defining essential concepts such as control points, contributors, and categorizations of such ecosystems. A clear understanding of these terms ensures alignment among stakeholders and prevents ambiguities in strategic decision-making.

Innovation ecosystems consist of interconnected players, each fulfilling a distinct role (the operating model according to Chap. 2 in this book). In the automotive industry, this includes OEMs, suppliers, software developers, regulatory bodies, and customers. These ecosystems operate through key control points, determining who contributes and who benefits, i.e., how value is created and distributed.

In the context of SDVs ecosystem design, these ecosystems are shaped by interdependent dynamics. First, the balance between competition and cooperation is essential ("coopetition," see Chap. 1 in this book). Unlike traditional supply chains with clearly delineated roles and hierarchical relationships, an SDV ecosystem operates through a fluid interplay of "coopetition." Companies must collaborate on shared foundations such as platforms and standards, while simultaneously competing through differentiated features and user experiences. Second, a wide range of contributors, from hardware suppliers and system integrators to cloud providers and app developers, must be coordinated across domains and organizational boundaries. Third, the ability to set standards and define integration methodologies becomes a source of strategic influence, impacting everything from interoperability to innovation speed.

As an SDV ecosystem matures (aligns), its evolution is driven by three overarching factors that shape collaboration and drive innovation. Herein standardization efforts (as a requirement for modularization) play a central role as a value driver (see Chap. 2 in this book) but also to establish de facto norms for interfaces, communication protocols, and safety frameworks (as part of the governance, see also Chap. 2). These standards are required for achieving seamless interoperability across different systems and suppliers. Following standardization, open-source initiatives promote transparency, reduce development redundancies, and accelerate technological progress. By lowering entry barriers and fostering collective innovation, they enable a broader range of contributors to participate meaningfully in the ecosystem. Last but not least, modular architectures ensure long-term scalability and flexibility. By decoupling hardware and software layers, they allow for faster iteration cycles, easier upgrades, and parallel development of features across organizational and geographical boundaries [2].

Against this backdrop, the automotive industry has undergone profound transformations, driven by continuous technological advancements. But yet, the current software-defined era to come is showing the progressive shift to increasingly software-centric vehicle architectures, demonstrating how developments have built upon the previous innovations, leading to the intelligent, connected vehicles of today. Figure 4.1 illustrates these transformation steps of the automobile from the mechanics-dominated era in the early 1900s, through the embedded software era of the 1990s, to the software-defined era. Initially, core mechanical components like the engine and drivetrain dominated vehicle design, requiring frequent manual maintenance and offering minimal automation with entirely mechanical controls. As vehicle systems grew more complex, electronics were introduced to provide precise control over mechanical components, with individual electronic control units (ECUs) managing tasks like fuel injection and the anti-lock braking system (ABS), improving safety and reliability while using sensors for diagnostics. In the latest phase, advanced AI and software govern critical functions, enabling autonomous driving and facilitating OTA updates, which enhance vehicle functionality and efficiency, transforming cars into “devices on wheels” and supporting new business models for continuous revenue.

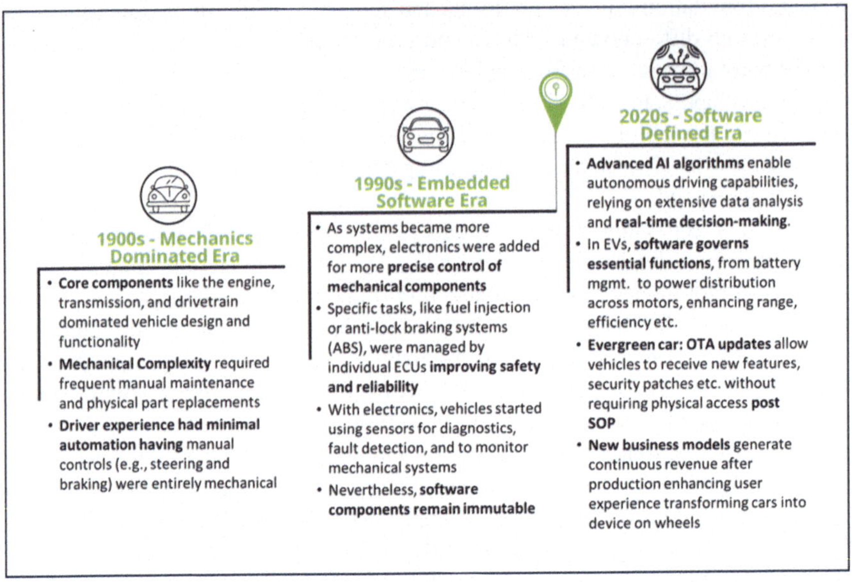

Fig. 4.1 Transformation steps of the automobile. *Source* Own design

Software is rapidly becoming the primary driver of innovation and value in the automotive industry, marking a shift from traditional hardware-centric development to SDVs. Today's OEMs are investing heavily in SDVs, with expenditures reaching up to US$3 billion per company annually, allocating substantial resources to research and development (R&D) [1], expecting 81% of the value contribution of the vehicles in an OEM's fleet to be software-driven by 2030 [1]. This shift transforms the industry from a traditional, fixed start of production (SOP)-driven approach to a cyclical hardware-software refresh model that ensures continuous innovation and technological advancement, in effect working toward an "evergreen" concept, where customers continuously receive OTA software updates, ensuring their vehicles remain technologically relevant long after SOP.

A key enabler of this transformation is the progressive decoupling of hardware and software, leading to higher integration and fewer components. Unlike legacy architectures, where software was deeply embedded within ECUs, modern SDV architectures make software more accessible and modular. This enables seamless backend management, allowing OEMs to deploy software bug fixes and security patches throughout the vehicle's lifecycle as well as introduce new features and functionality post-SOP, enhancing customer experience (see Chap. 3 in this book). This is expected to lead to the development of defacto standards in the automotive sector. Market-driven harmonization across interfaces and architectural layers will define interoperability, rather than rigid regulatory mandates.

The shift is not just about technology but also about redefining industry roles. As already mentioned, a thriving SDV ecosystem requires the participation of OEMs, suppliers, tech players, cloud and connectivity providers, regulatory bodies, and mobility service providers. Hyperscalers and telecom carriers will play pivotal roles in defining connectivity backbone of future vehicles. Meanwhile, insurance providers, fleet operators, and shared mobility orchestrators (Chap. 6 in this book) will reshape how vehicles are financed, insured, and utilized. They could be partners in ecosystems that are far more interconnected than traditional automotive supply chains – and far more dynamic. Success in the SDV era will require cross-industry collaboration, where diverse players contribute to an open, scalable, and interoperable SDV ecosystem. In the following, we provide a closer look at the key players involved.

At the center of this transformation are (still) the OEMs, evolving from hardware-centric development to orchestrating software-defined architectures built on centralized computing platforms. Their future competitiveness will depend on how effectively they are in this integrative role (see Chap. 1 in this book). This involves three strategic imperatives (see also Chap. 3 in this book):

1. Aligning to harmonized SDV architecture – to establish a unified foundation for software and hardware integration across partners.
2. Identifying and securing key technology suppliers – to build robust and adaptable software stacks that can evolve with changing demands.
3. Developing a strategic ecosystem roadmap – to define clear roles, responsibilities, and milestones to coordinate internal development with external collaboration.

Alongside the OEMs, other contributors are equally critical:

- Tier-1 suppliers are shifting from component manufacturers to providers of integrated hardware-software solutions, increasingly involved in software-defined component development.
- 3rd party developers leverage open-source platforms and service-oriented architectures (SOA) to deliver modular, scalable automotive applications that enhance vehicle functionality and user experience.
- Connectivity players enable real-time data exchange, vehicle to everything (V2X) communication, and secure OTA updates for continuous feature deployment and vehicle enhancement.
- Cloud providers provide the backbone for scalable computing, continuous integration/continuous deployment (CI/CD) pipelines, AI-powered analytics, and comprehensive data management across vehicle fleets.
- Fleet operators apply predictive maintenance, remote diagnostics, and cloud monitoring to improve efficiency and reduce lifecycle costs.

Together they all work in establishing the onboard and offboard hardware and software architecture able to accelerate digital features, lower cost, create seamless user experience and reach scale in a comprehensive open SDV ecosystem (Fig. 4.2).

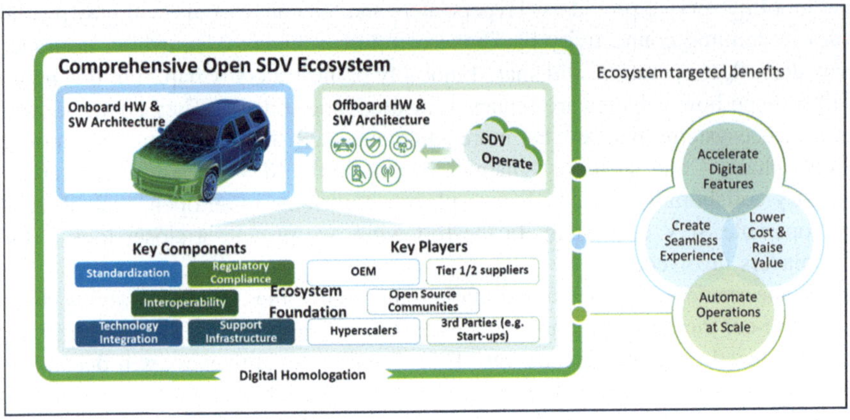

Fig. 4.2 A comprehensive open SDV ecosystem. *Source* Own design

As explained above and depicted in Fig. 4.2, a comprehensive and open future-proof automotive SDV ecosystem requires a deep understanding of three key components: value drivers (above all standardization and interoperability allowing modularization and scaling in network) as well as governance mechanisms (such as regulatory compliance as core of the effective governance) and also the platform (technology integration and support infrastructure), see Chaps. 2 and 3 in this book. Furthermore, such an ecosystem requires the already mentioned key players.

Such key components and players are aimed to automate operations at scale as an ecosystem-targeted benefit. Key components and key players also influence the hard and software. As ecosystem-targeted benefits, this impact relationship enables the creation of a seamless experience, lower costs and a higher (jointly created) value in the ecosystem.

The resulting hardware and software architecture is initially the onboard hardware and software architecture, which in turn influences the offboard hardware and software architecture, i.e., the SDV operating platform. This results in an acceleration of digital features as ecosystem-targeted benefits. In the following, the requirements for the onboard and the offboard vehicle architecture are examined in more detail (Sects. 4.3 and 4.4).

4.3 Requirements for the Onboard Vehicle Architecture: Software and Hardware

As the automotive industry transitions toward SDVs, the requirements placed on an onboard—hardware (HW) and software (SW)—vehicle architecture have become paramount. This transformation demands a robust combination of (1) high-performance computing (HPC) & centralized architecture, (2) scalability & modularity, (3) cybersecurity & functional safety, (4) software & OTA update capabilities as well as (5) sensor fusion & perception stacks (see Fig. 4.3).

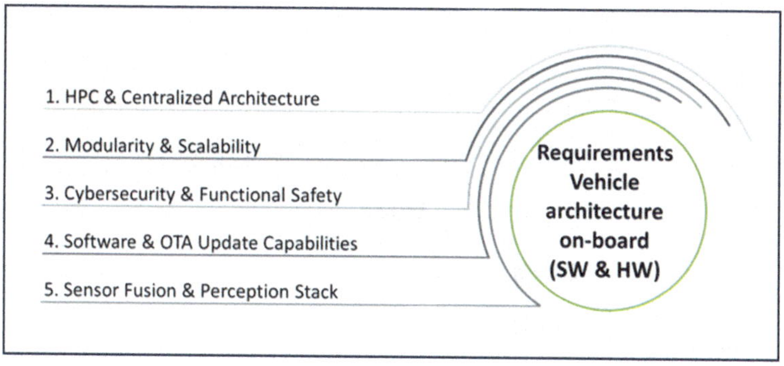

Fig. 4.3 Requirements placed on an onboard vehicle architecture. *Source* Own design

1. HPC & Centralized Architecture:

Previously, hardware-centric designs with distributed architecture required one or more ECUs to implement one feature in a vehicle. However, with the emergence of centralized architecture with HPC capabilities, features require the implementation of software functions either in-vehicle or in-cloud. Centralized HPC platforms are essential and pivotal in future vehicles for processing data from various sensors with low latency, enabling seamless advanced driver assistance systems (ADAS), infotainment, and connectivity features. OEMs are increasingly adopting centralized architectures (38% predicted market share by 2030) [1]. By consolidating multiple ECUs into a central computing unit, manufacturers can enhance processing efficiency, reduce wiring complexity, and improve energy consumption (see Fig. 4.4 for details).

In a distributed architecture, each function or feature in the vehicle has its own dedicated ECU. These ECUs are connected through a network, such as a Controller Area Network (CAN) bus. Each ECU controls its specific function, such as engine control, brake control, or entertainment system. This approach has been widely used due to its simplicity and modularity, but it can lead to a large number of ECUs in a vehicle, increasing complexity, weight, and cost.

A hybrid electrical/electronic (E/E) architecture combines domain-specific computing centralization with zonal cabling optimization. Domain control units (DCUs) bundle functionally related computing loads, while zone control units (ZCUs) act as local nodes that accommodate numerous ECUs, thereby minimizing cable lengths. Data traffic flows between the DCU layer and the ZCU layer via a uniform high-speed Ethernet that supports heterogeneous real-time and infotainment messages in parallel. This allows functions to be integrated flexibly, software to be scaled more easily, and hardware costs to be reduced.

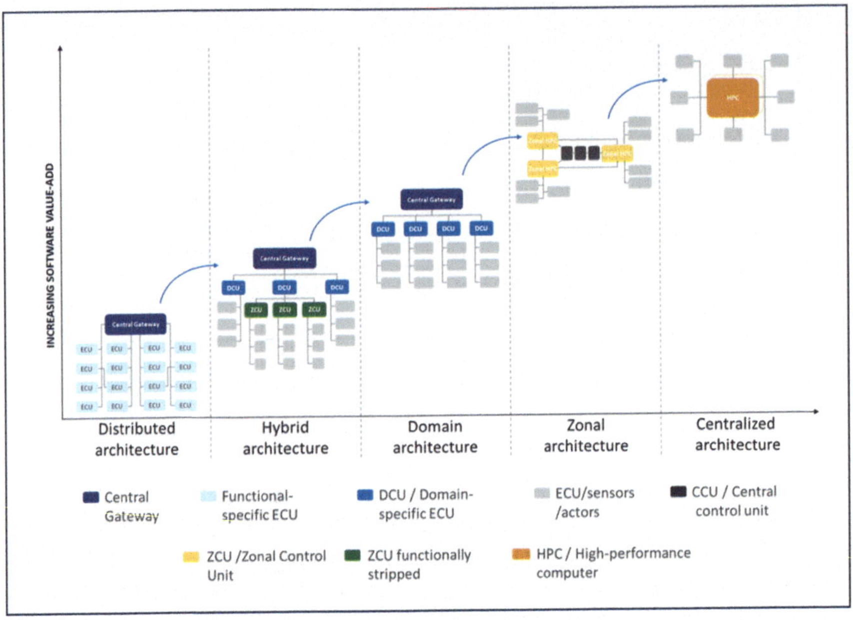

Fig. 4.4 Evolution of automotive onboard hard- and software architecture. *Source* Deloitte, 2024 [1]

In a domain architecture, the vehicle's functions are grouped into domains, such as powertrain, chassis, safety, and infotainment. Each domain has its own powerful computer that controls multiple functions within that domain. This approach allows for better integration of functions within a domain and can reduce the number of ECUs compared to a distributed architecture. However, like the zonal architecture, it requires more advanced networking technologies and can present new challenges in terms of system integration and safety.

In a zonal architecture, the vehicle is divided into zones, and each zone has a high-performance computer that controls multiple functions within that zone. This further reduces the number of ECUs and the complexity of the network. It also allows for more powerful computing capabilities and easier software updates. However, it requires more advanced networking technologies and can present new challenges in terms of system integration and safety.

In a centralized architecture, the single HPC acts as the central processing unit (CPU) for the vehicle, receiving inputs from various sensors and actuators, processing the data, and sending commands to the appropriate systems and components. In this approach, a single, powerful HPC manages all the functions and features of the vehicle. This approach can provide greater integration and coordination of the vehicle's functions but can also be more demanding on migrating software features and expensive. While in theory a single HPC is possible, we expect a back-up HPC component to be added to the architecture in order to guarantee redundancy in case of a safety–critical failure. Nonetheless, the centralized architecture leads to a drastic complexity reduction in the vehicle.

The integration of HPC in vehicles supports the increasingly demanding processing requirements, enabling a feature-rich platform that can endure for a vehicle's lifetime. These systems require immense computational power, often provided by specialized hardware accelerators such as graphics processing units (GPUs), neural processing units (NPUs), and field-programmable gate arrays (NPGA). The ability to decouple software from hardware is crucial, allowing software development, deployment, and upgrades to occur independently of hardware limitations, which is particularly important in the automotive industry where multiple third parties provide software for OEMs [3].

The evolution of ADAS and the vehicle's E/E architectures toward SDV paradigms emphasizes the role of centralized computing and AI-specific hardware. Figure 4.5 mirrors this as by 2030, centralized systems are predicted to dominate. The industry has been working on addressing these challenges by developing a proof of concept that emphasizes the decoupling of software from hardware. Different initiatives aim to revolutionize vehicle development, enabling faster hardware independent software updates, enhanced cybersecurity, and more efficient long-term maintenance. By embracing centralized computing architectures, the automotive industry is paving the way for smarter, safer, and more adaptable vehicles that remain technologically relevant throughout their lifecycle.

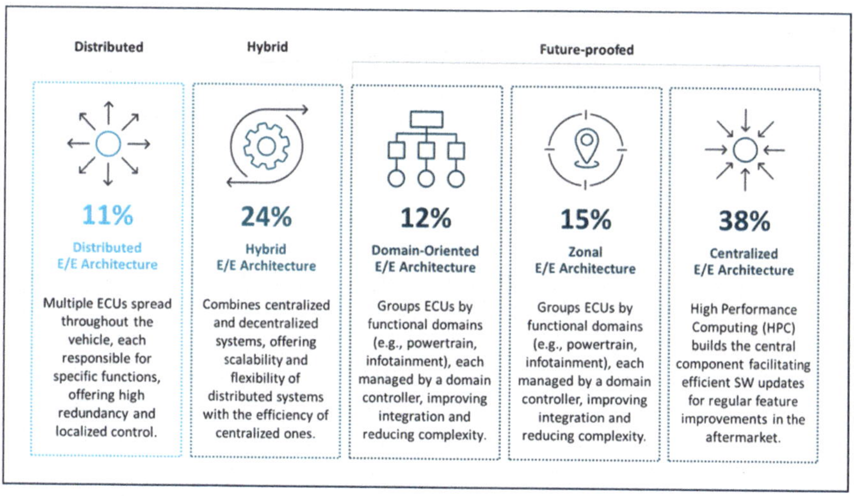

Fig. 4.5 Predominant E/E architecture in 2030. *Source* Deloitte, 2024 [1]

2. Modularity & Scalability

Modularity and scalability are crucial aspects of SDVs, enabling them to adapt to evolving technologies and market demands. A flexible architecture that can accommodate continuous updates and expansions without extensive hardware modifications are essential. This is achieved through a modular design that decouples software components, allowing them to be developed, updated, and integrated independently [4]. By structuring software into distinct modules, SDVs enhance scalability, making it easier to add or remove features without overhauling the entire system. This modular approach also improves interoperability, enabling different modules to communicate and exchange data efficiently in real-time. Additionally, it streamlines the integration of new functionalities, ensuring smooth enhancements without introducing additional layers of code [4].

A key enabler of scalability in SDVs is the adoption of a SOA. SOA organizes software into discrete services, each encapsulating a specific functionality. This approach supports interoperability across different platforms and programming languages while enabling the reuse of existing services to build more complex applications [5]. By leveraging SOA, manufacturers can accelerate development cycles, facilitate rapid prototyping, and enhance overall system efficiency.

Furthermore, scalable SDV architecture is designed to anticipate future advancements, allowing new software features to be integrated seamlessly without disrupting existing systems. This forward-thinking approach ensures that vehicles remain technologically relevant throughout their lifespan, adapting effortlessly to emerging innovations [6].

As a result, modular SDV architecture contributes to reduced complexity, enabling streamlined development processes, improved maintainability, and enhanced scalability.

3. Cybersecurity & Functional Safety

As vehicles become increasingly software-driven, cybersecurity and functional safety have emerged as critical pillars of modern automotive development. Given that vehicles remain in operation for 15 years or more, manufacturers must ensure that their software not only meets current security standards but also remains resilient against emerging threats over time. Cyberattack methods that do not exist today may become a significant concern in the future, requiring proactive and continuous protection throughout a vehicle's lifecycle.

One of the key trends in automotive cybersecurity is the automation of security workflows. Organizations are leveraging automated security processes and machine learning models to improve threat detection accuracy, reduce false positives, and enhance response times. This shift has led to a notable reduction in the mean time to respond (MTTR) to security incidents, emphasizing the importance of real-time, adaptive defense mechanisms [7]. At the same time, functional safety remains a foundational aspect of SDVs, ensuring that systems continue to operate reliably despite E/E malfunctions. Certification of functional safety measures allows manufacturers to meet stringent industry safety standards.

The convergence of cybersecurity and functional safety [8] is now playing an increasingly important role in optimizing both vehicle security and operational reliability. Addressing these two critical areas at the system level enables manufacturers to comply with evolving regulations while also gaining a competitive edge – improving time to market and enhancing vehicle resilience. This growing need for long-term security and compliance brings us to the concept of the already mentioned OTA updates. As cybersecurity threats evolve, manufacturers must ensure that vehicles remain protected throughout their lifespan. Regulatory bodies now mandate that automakers guarantee long-term cybersecurity, making periodic software updates a legal obligation rather than an optional feature. OTA updates provide a crucial mechanism for delivering these ongoing security enhancements, ensuring compliance with new regulations and safeguarding vehicles against emerging cyber threats.

Ultimately, integrating robust security and functional safety measures from the outset allows manufacturers to build vehicles that offer greater trust, performance, and longevity. In an ever-evolving automotive landscape, a dynamic cybersecurity strategy – supported by automation, machine learning, and OTA updates – will be essential to maintaining the safety, reliability, and regulatory compliance of modern vehicles throughout their extended operational life.

4. Software & OTA Update Capabilities

OTA updates have become a cornerstone of modern automotive development, enabling manufacturers to remotely enhance vehicle functionality, fix software bugs, and introduce new features throughout a vehicle's lifecycle. As the scope and complexity of OTA capabilities grow, so does the need to ensure that safety–critical and non-safety–critical systems are handled with appropriate levels of rigor, oversight, and security.

The separation of OTA capabilities by safety relevance is a crucial architectural and procedural principle designed to protect the integrity of safety–critical vehicle functions while still allowing the agility and flexibility needed for non-critical updates. This separation ensures compliance with functional safety standards (such as ISO 26262), minimizes the risk of unintended interactions, and supports cybersecurity goals (Fig. 4.6).

In Fig. 4.6, OTA functions are categorized in the two main domains: safety relevant and non-safety relevant. Safety relevant capabilities include updates to critical subsystems such as cybersecurity, ADAS, and powertrain controls. These updates are directly tied to the safety and security of passengers and require high assurance in both design and deployment. On the other hand, non-safety relevant capabilities relate to enhancing the driving experience, comfort, and convenience, covering updates to infotainment, climate control, and keyless entry.

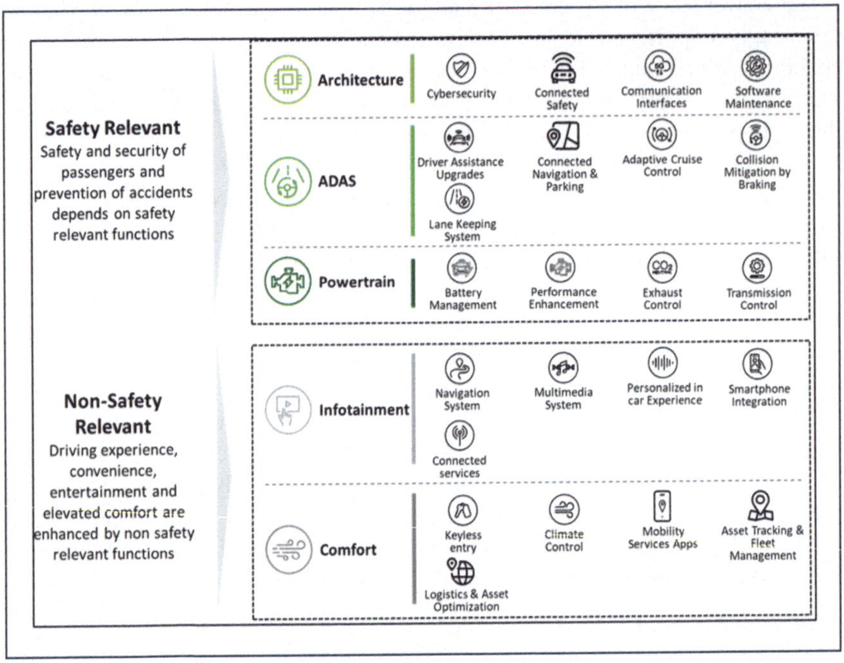

Fig. 4.6 Separation of OTA capabilities by safety relevance. *Source* Own design

The integration of software and OTA update capabilities is driving rapid growth in the global automotive OTA market, projected to increase from US$3.3 billion in 2022 to nearly US$14.0 billion by 2030 [9]. Figure 4.7 highlights the effectiveness of OTA updates according to the cost savings expected in a global manufacturer readiness study [1] if faulty software could be corrected via OTA update. Herein over 160 executives from OEMs in Germany, France, Italy, United Kingdom, Spain, the United States, Japan, and South Korea were interviewed from March to April 2024 to gather insights on the rapidly evolving nature of SDVs and the transformational impact this trend is having on automotive manufacturing organizations around the world. Regarding the avoidance of recalls, OTA updates do not only enhance vehicle reliability and customer satisfaction but also lead to significant cost savings. Traditional recalls involve substantial expenses related to logistics, labor, and replacement parts, as well as inconvenience for vehicle owners. In contrast, OTA updates allow manufacturers to deploy fixes swiftly and efficiently, minimizing financial burdens and ensuring a seamless user experience.

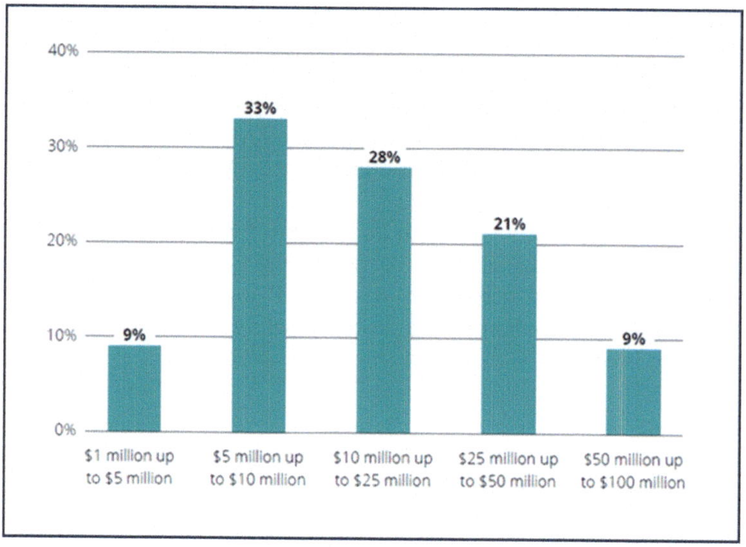

Fig. 4.7 Cost savings if faulty software could be rectified via OTA update. *Source* Deloitte, 2024 [1]

However, the implementation of OTA updates is subject to regulatory scrutiny, as seen in China's mandate for regulatory approvals for autonomous driving-related software upgrades. This regulation aims to prevent OEMs from using OTA updates to conceal defects or avoid liability, requiring them to obtain licenses for product changes before resuming production of affected vehicles [10]. Such regulatory measures ensure that OTA updates are conducted responsibly, maintaining safety and transparency in the automotive industry.

5. Sensor Fusion & Perception Stack

Sensor (data) fusion and perception stacks are integral components of SDVs, enabling them to interpret complex environments with high accuracy and efficiency. Sensor fusion involves combining data from multiple sensors, such as radars, lidars, and cameras, to create a comprehensive model of the vehicle's surroundings. This process reduces uncertainty and enhances the reliability of the information used for decision-making.

The perception stack in SDVs utilizes sensor fusion to support intelligent actions by balancing the strengths and weaknesses of different sensor modalities. For instance, radars excel in determining distance and speed even in adverse weather conditions, while cameras are adept at reading signs and classifying objects like pedestrians and vehicles. Lidars provide precise object detection but lack the range and affordability of cameras or radars. By integrating data from these diverse sensors, SDVs can achieve a more accurate and reliable representation of their environment.

Advanced sensor fusion techniques, such as low-level sensor fusion, require a high processing power that is centralized in a designated ADAS domain controller, allowing for real-time data collection and fusion from each sensor. This approach reduces latency and enhances performance, enabling vehicles to make rapid decisions in critical situations. Additionally, the use of AI tools in sensor fusion can extract valuable information from sensor data, improving detection probability and reliability.

In summary, a future-proof onboard architecture for SDVs must combine centralized computing, modular software design, robust cybersecurity, seamless OTA capabilities, and advanced perception systems, all together enabling continuous innovation, reducing complexity, and ensuring safety and scalability across the vehicle's entire lifecycle.

4.4 Requirements for Offboard Vehicle Architecture: Software and Hardware

A well-structured offboard architecture ensures seamless connectivity, secured data handling, and efficient OTA updates. As shown in Fig. 4.8, requirements include (1) cloud computing & edge infrastructure, (2) an OTA update management system, (3) digital services & connectivity, and (4) AI-driven analytics, each of which will be explained in the following.

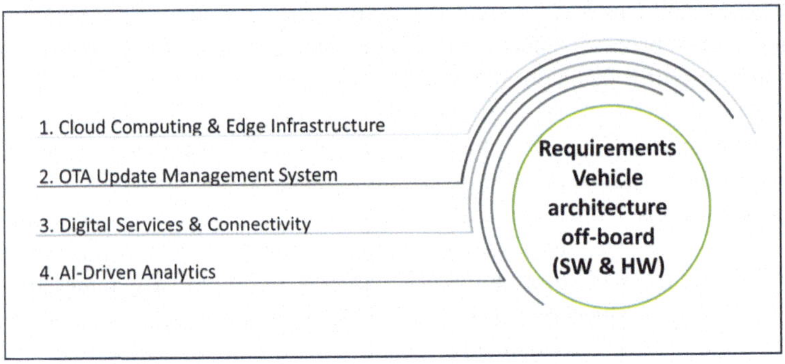

Fig. 4.8 Requirements placed on an offboard vehicle architecture. *Source* Own design

1. Cloud Computing & Edge Infrastructure
   Cloud computing has expanded its role beyond mere storage and computation, evolving into a crucial enabler for SDVs. Through its ability to provide agility, scalability, and intelligence, cloud computing supports various automotive applications, from data management to software deployment. In this context, the shift toward multi-cloud architectures gives organizations the flexibility to leverage different cloud providers. This multi-cloud strategy offers benefits such as enhanced performance, operational continuity, and risk mitigation, making it easier for automotive manufacturers to manage global operations and ensure business resilience.
   Meanwhile, edge infrastructure complements cloud computing by decentralizing data processing closer to the source at the edge of the network, which helps reduce latency. This is particularly important in applications like autonomous driving and real-time traffic management, where quick processing and decision-making are critical. Edge computing allows for faster local data processing, enabling immediate responses to dynamic conditions in the vehicle's environment without waiting for instructions from the cloud. This decentralized approach is essential for delivering real-time capabilities, ensuring that the vehicle systems can act on data as soon as it is gathered.

2. OTA Update Management System
   One of the key innovations enabled by cloud computing and edge infrastructure is the backend system that supports OTA update management, a critical component for SDVs (see also Chap. 3 in this book). To guarantee seamless, secure, and efficient software updates, the backend system must meet several stringent operational and technical requirements as seen in Fig. 4.9.

Fig. 4.9 Key requirements for a robust OTA backend infrastructure. *Source* Own design

To effectively support the dynamic and continuous demands of OTA update management, a robust OTA backend infrastructure must be built upon several critical pillars. First and foremost, it must ensure 24/7 operations and high availability, enabling uninterrupted software delivery across global vehicle fleets, regardless of time zone or network conditions. The infrastructure must be globally distributed, leveraging regional cloud instances and Content Delivery Networks (CDNs) to guarantee worldwide reachability and consistent performance across geographic locations. To reduce latency and enhance responsiveness, especially in safety–critical applications, the system must support low latency and fast data processing through edge computing capabilities, allowing updates to be cached or processed closer to the vehicles themselves.

Another essential requirement is interoperability and modular architecture. This means the backend must seamlessly integrate with a variety of in-vehicle platforms and ECUs via standardized APIs and protocols. This flexibility enables coordination between OEMs, Tier-1 suppliers, and third-party service providers, ensuring consistent behavior across different hardware and software environments. Security and compliance are equally critical, given the safety-sensitive nature of automotive systems. OTA systems must implement advanced encryption, authentication, and cybersecurity mechanisms in alignment with international standards like ISO/SAE 21,434, protecting against unauthorized access and tampering.

Moreover, the infrastructure must feature efficient bandwidth management to adapt to diverse network conditions – such as Wi-Fi or cellular – by enabling partial downloads, smart compression, and the ability to resume interrupted updates. This ensures reliable performance even in constrained connectivity environments. Lastly, data integrity and version control are vital to maintaining system stability. The backend must accurately track software versions, dependencies, and rollback mechanisms to avoid conflicts and preserve consistency throughout the vehicle lifecycle.

When all these elements are effectively implemented, the OTA infrastructure provides a secure, scalable, and resilient foundation for managing software across the vehicle fleet. This not only improves vehicle performance, safety, and user experience, but also empowers automotive manufacturers to deliver continuous innovation while minimizing operational disruptions and maintenance costs.

3. Digital Services & Connectivity

The automotive sector is increasingly focusing on digital services and enhanced connectivity to enhance driving safety and efficiency. These services enable real-time interaction between vehicles, infrastructure, and other connected devices, creating a more responsive and adaptive driving environment. For instance, communication between vehicles and infrastructure, such as traffic lights and road sensors, can optimize traffic flow and reduce congestion. By leveraging data from connected vehicles, cities can manage traffic in smarter ways, improving overall road safety and efficiency.

4. AI-Driven Analytics

AI-driven analytics are crucial in managing and analyzing the vast amounts of data generated by SDVs. These vehicles are equipped with sensors, cameras, radar, and other devices that produce massive amounts of real-time data. AI can be leveraged to analyze these data streams for a variety of purposes, from identifying anomalies that suggest potential cyberattacks to optimizing driving patterns and predicting vehicle maintenance needs [11]. The integration of AI enhances the processing of data by enabling predictive analytics and real-time decision-making. For instance, AI can process data from vehicle sensors in the cloud, providing insights that improve route planning or traffic management. Furthermore, AI can help vehicles make decisions on the fly, such as avoiding obstacles or adapting to changing traffic conditions, which is crucial for the development of autonomous driving. However, AI computing is not solely dependent on cloud infrastructure, many manufacturers are also integrating AI processing directly into the vehicle itself. Edge AI computing allows SDVs to perform critical real-time analysis locally, reducing latency and ensuring safety–critical decisions, such as obstacle avoidance and adaptive cruise control, are made instantly without relying on cloud connectivity.

By distributing AI processing across both cloud and onboard computing systems, SDVs can optimize performance while maintaining the ability to function independently when network connectivity is limited. AI-powered in-vehicle processing is especially vital for autonomous driving, where real-time perception, sensor fusion, and decision-making must happen within milliseconds. Additionally, onboard AI computing helps improve cybersecurity by detecting threats locally before they propagate through the vehicle network.

Cloud platforms also play a crucial role in AI analytics as they enable the management and analysis of this enormous volume of data. This is because they turn raw information into actionable insights that improve both, the driver experience and the vehicle's performance. Additionally, the use of AI allows for the automation of many tasks, improving efficiency and reducing the potential for human error in vehicle operation.

4.5 Importance of Scaling SDV Ecosystems

Leading digital ecosystems, such as Apple's App Store, Amazon Web Services (AWS), and Google's Android, demonstrate the power of scalability and network effects as a value driver (see Chap. 2 in this book). Open platforms that allow third-party integration tend to attract more developers and foster innovation at scale. All these cloud-driven platforms enable real-time updates, predictive analytics, and enhanced customer experiences, i.e., an overall value proposition.

Translated to the SDV ecosystem this means that a large ecosystem enables faster software iterations, ensuring continuous improvement and higher performance. Greater ecosystem participation by OEMs, software providers, and startups can commonly create a self-sustaining innovation cycle. Beyond this scaling that operates backend across multiple OEMs ensures cost efficiency in hardware and software development, reducing duplication efforts.

How connected vehicle fleets can be integrated and managed are crucial in context of scale. As automotive manufacturers expand their ecosystems, they can connect a larger number of vehicles, enhancing fleet capabilities and the value derived from data. A larger fleet improves network robustness, enabling better data collection, real-time monitoring, and predictive maintenance, which enhance performance and customer satisfaction. Furthermore, scaling facilitates the integration of new technologies, such as autonomous driving, by providing a broader base for testing and refining autonomous systems. Cooperation with partners in the ecosystem is important for all elements of the vehicle (software and hardware) architecture – both onboard and offboard [2].

Scale itself is a fundamental driver of success in the SDV ecosystem and occurs across three dimensions (see Fig. 4.10).

Scaling a digital ecosystem broadens its geographic reach, ensuring consistent and reliable services across the globe. Manufacturers can offer region-specific features such as localized traffic updates, language preferences, and compliance with local regulations. Global scaling provides manufacturers with economies of scale, reducing operational costs and enhancing service delivery efficiency. It also facilitates the collection of data from diverse regions, offering deeper insights into consumer behavior, which fuels innovation and improves services.

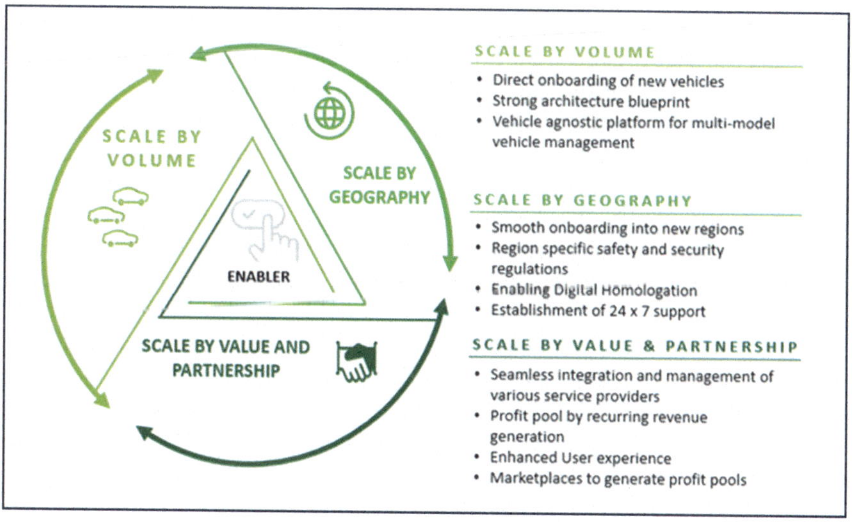

Fig. 4.10 Power of scale–Scaling by volume, geography, and value & partnership. *Source* Deloitte, 2024 [1]

Significant efficiencies can be also achieved by deploying a unified architecture across a wide range of vehicles without redesigning the core system for each model e.g., by decoupling the software stack from the underlying vehicle. A platform agnostic approach to architecture supports high volume scalability beyond models, brands and even OEMs by leveraging standardized components, centralized computing and a common software foundation that is flexible enough to be configured to various vehicle segments. As a result OEMs can roll-out updates OTA, introduce new features post-sale and deliver consistent digital user experience at industrial scale.

Developing an SDV app store would allow developers to contribute software solutions and expand an SDV ecosystem with a wide range of offerings—from basic connectivity to advanced features like autonomous driving (scale by value and partnerships). As the ecosystems grow, manufacturers can offer personalized in-car experiences, remote diagnostics, and software updates, while also generating new revenue through subscription-based services. Furthermore, scaling enables partnerships with industries such as telecommunications, insurance, and transportation, fostering new business models like integrated mobility solutions and dynamic pricing. These opportunities help automotive companies diversify their services and strengthen their competitive position.

4.6 Benefits of the SDV Ecosystems in Automotive - An Outlook

The establishment of a strong digital and operational SDV ecosystems in the automotive industry brings significant benefits, including lower development costs, higher innovation potential when open, and quality improvements – but only if robust quality gates are in place, as openness alone does not automatically ensure higher quality. These factors are critical for fostering innovation, sustainability, and resilience in an industry undergoing rapid transformation.

A mature SDV ecosystem enhances efficiency primarily through an effective development toolchain and well-structured teams. Standardization plays a crucial role in achieving these efficiencies by ensuring compatibility, reducing integration efforts, and streamlining development processes. By adopting standardized software architectures and interfaces as well as open-source frameworks, companies can reduce redundancies, improve interoperability, and accelerate development cycles.

Aligned with broader sustainability goals, SDVs offer new opportunities to implement circular economy principles in practice. Central to this vision is the concept of the "evergreen car," which enables vehicles to remain technologically relevant over their entire lifecycle through continuous software updates. By decoupling innovation from hardware replacement cycles, OEMs can reduce material waste, avoid early obsolescence, and lower lifecycle emissions. This shift has significant implications not only for the environment but also for the value proposition of SDVs. Used vehicles that receive regular OTA updates retain greater functional parity with newer models and can therefore command higher resale values. This prolongs the usable lifespan of mobility assets and supports a more circular use of resources. Furthermore, the ability to update vehicles post-sales strengthens the link between sustainability and user satisfaction. Instead of being left behind, customers benefit from ongoing feature improvements and enhanced safety without the need to purchase a new vehicle [12].

Furthermore, digital technologies, including AI-driven automation and real-time data exchange, significantly boost operational efficiency of SDV ecosystems. Automated workflows reduce manual interventions, improve decision-making speed, and enhance supply chain visibility. Efficiency gains through the adoption of SDVs are estimated at up to 20%, particularly in software development, testing, and integration processing [1]. This digital integration ensures that manufacturers can swiftly adapt to disruptions, optimize logistics, and maintain a competitive edge in a fast-evolving market.

Beyond technical harmonization, partnerships also enable faster and more efficient development cycles. Through a parallel division of labor, ecosystem partners can work simultaneously on specialized subsystems, leveraging their respective core competencies. This approach not only accelerates time-to-market but also facilitates resources and risk sharing, particularly in areas involving high upfront R&D investments or regulatory uncertainty such as SDV. The Deloitte SDV Readiness Study [1] quantifies the benefits of such partnerships: organizations that are already actively in collaborative ecosystems report R&D cost savings of up to 16% compared to those operating in isolation. These savings stem from shared R&D and innovation costs, access to broader expertise, reduced time to market due to a streamlined integration processes, and more efficient use of development resources as well as shared software platforms in innovation ecosystems:

- Shared R&D costs: Developing the software stack, middleware, and new E/E architectures for SDVs is capital-intensive, while a joint development (e.g., through alliances or co-engineering platforms) spreads R&D investment across partners, reducing individual OEM outlays.
- Reduced time-to-market: Partnerships accelerate product development through resource pooling and faster learning cycles. Shorter timelines mean earlier revenue capture and fewer delays, resulting in cost avoidance.
- Economies of scale in procurement: Partnering OEMs can align on key hardware/software components (e.g., chips, sensors, ECUs), enabling bulk purchasing, vendor leverage, and lower per-unit costs.
- Shared software platforms: A common, modular software platform reduces duplication and enables reuse across models and brands lowers integration and maintenance efforts.
- Fewer warranty costs & field failures: Higher software quality through shared testing infrastructure and validation environments leads to lower post-sale costs and increased customer satisfaction.

These savings compound over time in an SDV context where ongoing updates, new feature development, and system validation are continuous efforts. Participating in SDV innovation ecosystems help OEMs move faster while staying cost-efficient.

In summary, SDV innovation ecosystems unlock significant economic and innovation advantages. This is possible through scale, collaboration, and continuous development, positioning OEMs to remain competitive in a rapidly evolving mobility landscape.

References

1. Deloitte. (2024). *Global manufacturer readiness study SDV*. Retrieved May 2, 2025, from https://www.deloitte.com/et/en/Industries/automotive/analysis/software-defined-vehicles.html
2. Liu, Z., Zhang, W., & Zhao, F. (2022). Impact, challenges and prospect of software-defined vehicles. *Automotive Innovation, 5*(2), 180–194. https://doi.org/10.1007/s42154-022-00179-z
3. Küfen, J., Burkhard, C., Herjürgen, N., Dawan, A. (2023). Approaches and potentials of a future zonal hardware architecture. *ATZ Electronics Worldwide, 18,* 18–22. https://doi.org/10.1007/s38314-022-1433-5
4. Chandel, A. (2023). *AUTOSAR modularity and configurability*. LHP Engineering Solutions. Retrieved May 2, 2025, from https://www.lhpes.com/blog/autosar-modularity-and-configurability
5. Zacchi, J.-V., Clementi, E., & Mata, N. (2025). *APIKS: A modular ROS2 framework for rapid prototyping and validation of automated driving systems*. https://doi.org/10.48550/arXiv.2502.20507
6. Dhamsania, N. (2024). Software defined vehicles: A complete guide. Retrieved May 2, 2025, from https://intuions.com/blog/software-defined-vehicles/
7. Zhukabayeva, T., Othman, M., Adamova, A., & Zhukabayeva, T. (2025). Cybersecurity solutions for Industrial Internet of Things–Edge computing integration: Challenges, threats, and future directions. *Sensors, 25*(1), Article 213. https://doi.org/10.3390/s25010213
8. Alexiades, N. (2024). *Where cybersecurity meets functional safety*. Control Design. Retrieved May 2, 2025, from https://www.controldesign.com/connections/cybersecurity/article/55137964/where-cybersecurity-meets-functional-safety
9. Deloitte. (2023). *Software-defined vehicles: Engineering the mobility revolution*. Retrieved May 2, 2025, from https://www.deloitte.com/content/dam/assets-zone1/jp/ja/docs/industries/consumer/2025/jp-aut-the-software-defined-vehicle-revolution-en.pdf
10. Harvey, J., & Fletcher, P. (2025). *China mandates regulatory approvals for autonomous driving software upgrades*. Reuters. Retrieved May 2, 2025, from https://www.reuters.com/business/autos-transportation/china-mandates-regulatory-approvals-autonomous-driving-software-upgrades-2025-02-28/
11. Schmelzer, R. (2025). *AI takes the wheel in accelerating the automotive industry*. Forbes. Retrieved May 2, 2025, from https://www.forbes.com/sites/ronschmelzer/2025/02/27/ai-takes-the-wheel-in-accelerating-the-automotive-industry/
12. Guissouma, H. (2024). *Positionspapier: Chancen durch Standardisierung und Open Source Software im Mobilitätssektor*. KIT und Universität Stuttgart. https://publikationen.bibliothek.kit.edu/1000168320/152362892.

Future Challenges of Homologation and Opportunities for Business Building in Transition to SDVs as well as the Role of Traditional OEMs

5

Elmar Pritsch, Erik Thorwirth, and Corinna Cruceru-Weisbrod

Contents

Abstract

If the market launch of SDV is to be accelerated, the software must be constantly updated. This applies first of all to the safety-relevant functions, which include not only ADAS and the powertrain but also the technical architecture. They ensure the safety and security of passengers and prevent accidents. However, this also applies to non-safety-relevant functions such as infotainment and comfort, which are responsible for the car and driving experience and without which survival in the stagnating global automotive market is not possible.

E. Pritsch (✉) · E. Thorwirth
Deloitte Consulting Germany, Düsseldorf, Germany
e-mail: epritsch@deloitte.de

E. Thorwirth
e-mail: ethorwirth@deloitte.de

C. Cruceru-Weisbrod
Deloitte Consulting Germany, München, Germany
e-mail: CCruceru@deloitte.de

H. Proff et al. (eds.), *Future of the Automotive Industry*, Future of Business and Finance, https://doi.org/10.1007/978-3-032-26308-7_5

5.1 Introduction

If the market launch of SDV (see Chap. 4 in this book) is to be accelerated, the software must be constantly updated. This applies first of all to the safety-relevant functions, which include not only ADAS and the powertrain but also the technical architecture. According to Fig. 4.6 in Chap. 4, they ensure the safety and security of passengers and prevent accidents. However, this also applies to non-safety-relevant functions such as infotainment and comfort, which are responsible for the car and driving experience and without which survival in the stagnating global automotive market is not possible (see Fig. 4.6).

However, the permanent updates especially to the technical architecture must always comply with the regulatory framework. Otherwise, the vehicles would no longer be insured at all, because damage resulting from changes to the vehicle would then be at the expenses of the previously concluded insurance policies. This chapter takes a closer look at the design of the technical architecture in SDV (innovation) ecosystems and, in particular, the challenges of digital homologation, i.e., the granting of an operating license by an official authority.

The chapter is therefore structured as follows: First, one of the (technical) challenges in the transition to the SDV that need to be solved in innovation ecosystems will be highlighted – the homologation (Sect. 5.2): the problems due to the insurance relevance of SDVs, digital homologation as a prerequisite for accelerating the market launch of SDVs, requirements for the technical architecture in SDV ecosystems to guarantee permanent homologation and a comparison of the SDV architecture aspects in SDV ecosystems with smartphone and computer ecosystems. However, we will not forget the opportunities that will drive the transformation to SDV despite all the problems (Sect. 5.3) and discuss the possible future role of traditional automotive companies in transition to SDVs (Sect. 5.4).

5.2 Challenges of Homologation in Transition to SDVs

5.2.1 Challenges Due to the Insurance Relevance of SDVs

As vehicles become increasingly autonomous, questions surrounding liability in the event of fault or accident have gained unprecedented complexity leading to evolving risk perceptions and insurance demand for AVs [1]. A critical timing represents especially the handover phase between autonomous driving and manual control and the transfer of responsibility for the vehicle driving mode and any effects. If for instance an accident occurs during or immediately after this transition, it needs to be clear whether the driver or the system is responsible and who will be held liable in the event of fault. This ambiguity creates significant challenges for insurance providers, who must determine responsibility and allocate damages accordingly.

The difficulty is compounded by the fact that current legal and regulatory frameworks are not fully adapted to dynamic handover scenarios. Traditional insurance models assume clear driver control, yet in SDVs, responsibility may shift back and forth between human and machine within milliseconds. Without robust evidence of the system's status at the time of an incident, disputes over liability can delay claims and increase legal uncertainty for both insurers and consumers [2].

As driving responsibilities increasingly shift from humans to machines, insurers will need to consider factors like software reliability and system failures, which will likely lead to new forms of coverage for drivers, software developers, and hardware manufacturers. The liability for accidents is anticipated to shift toward OEMs and system providers, potentially affecting commercial auto premiums and expanding the role of reinsurers. Traditional advantages of insurers, such as brand recognition and acquisition expertise, may be inadequate as data becomes central in risk selection and claims management. As OEMs explore direct insurance sales and partnerships, they might become significant competitors. While some OEMs are developing full-stack insurance models, most will likely pursue partnerships to leverage insurance leads and share vehicle data.

At the same time, SDVs offer the potential to improve insurance processes through enhanced trackability and transparency. Modern vehicles are equipped with a vast array of sensors connectivity features that can provide detailed event logs, including speed, driver input, sensor readings, and system status, before, during, and after a critical incident. These digital footprints enable forensic-grade analysis and help reconstruct the exact sequence of events, making liability assessments more objective and accurate. Beyond accident reconstruction, SDV sensors support new types of proactive services. For example, after detecting a collision, the vehicle can autonomously assess the damage, identify the affected components, and place an order for the required spare parts directly from an authorized workshop. This capability not only reduces downtime and manual intervention but also strengthens the connection between mobility, maintenance, and insurance services. Insurance distribution is also on the verge of significant transformation.

These developments highlight that the insurance viability of SDVs is closely tied to their regulatory robustness. Establishing mechanisms for continuous traceability, data certification, and digital compliance do not only support insurers but also lay the foundation for dynamic homologation processes, topics further detailed in the following section.

Finally, we have noticed in the past the significant positive impact on safety introduced by more assisted driving features like ABS, ESP, lane keep assist (LKA) and collision avoidance (CA) reducing the human inattentiveness as the weak element in the safety chain. We expect that the higher penetration of SDV with the capability of stepping up from autonomous driving level 2 + assisted drive features to autonomous driving level 3 and 4 will go hand in hand with a reduced number of accidents [3]. The future will reveal if insurers benefit from the increase in safety that might be offset by a higher utilization of an autonomous SDV.

5.2.2 Digital Homologation as a Prerequisite for Accelerating the Market Launch of SDVs

The concept of type approval, namely vehicle homologation (technical approval) in the automotive industry, is undergoing a major transformation, shifting from static compliance processes to a more dynamic certification approach. Traditionally, homologation requires physical testing and certification of vehicles, ensuring that they meet safety, environmental, and regulatory standards before being sold. However, with the advent of SDVs, this approach is no longer sufficient. Digital homologation now plays a critical role in ensuring continuous compliance, particularly as OTA updates can directly impact safety–critical software [4].

One of the primary challenges of digital homologation lies in the need for continuous compliance monitoring. In an SDV, software updates, delivered via OTA, can modify the behavior of safety–critical functions, such as ADAS and AD features. Regulatory authorities mandate that any changes to these systems must be documented and re-certified to ensure they continue to meet safety and legal requirements. This ongoing re-homologation process ensures that every update maintains the integrity of safety–critical functions throughout the vehicle's lifecycle, rather than relying solely on a one-time certification before the vehicle is put on the road.

Rather than seeing the increasing complexity of this system as negative, we should view the new technologies, like AI and ML, as key parts of the solution. Especially for software developers, AI technologies help bridge the gap toward legal requirements addressing the complexities and bring new, innovative ways to reduce risk, enhance performance, and ensure ongoing compliance. The challenge is not that AI and ML make things harder, but that they require new approaches to validation, ensuring their safety and predictability in real-world conditions. These tools enable us to manage and optimize safety and regulatory compliance across a vehicle's entire lifecycle, which would be far more challenging without their capabilities.

New regulatory standards are emerging to guide the safe deployment of AI in vehicles. Ensuring compliance with these standards involves regular vehicle reassessments to validate that it operates safely and remains compliant with evolving regulations. This includes rigorous testing, certification, and ongoing monitoring after vehicles have been deployed. In addition to AI concerns, the concept of software lifecycle compliance is becoming increasingly important. Particularly for vehicles incorporating ASIL-D software (Automotive Safety Integrity Level D, the highest level of safety integrity), updates and changes to software during the vehicle's operational lifecycle must be carefully managed. Any changes to ASIL-D software require recertification to ensure that safety and compliance standards are still being met. This means that manufacturers must have robust systems in place to track software changes, assess their impact on vehicle safety, and ensure that updated software remains compliant with both safety standards and regulatory requirements.[1]

The lifecycle of a vehicle spans multiple stages, from concept development through to decommissioning (see Fig. 5.1). Traditional homologation (version 1.0) focuses on the vehicle itself, certifying that it meets safety and regulatory requirements at a single point in time – typically, the moment the vehicle enters production. The new approach to homologation (version 2.0), which focuses on SDVs, must account for software updates and functional changes that happen throughout the vehicle's lifecycle. This takes into account that homologation is no longer a one-time certification event but an ongoing process that spans across the following phases:

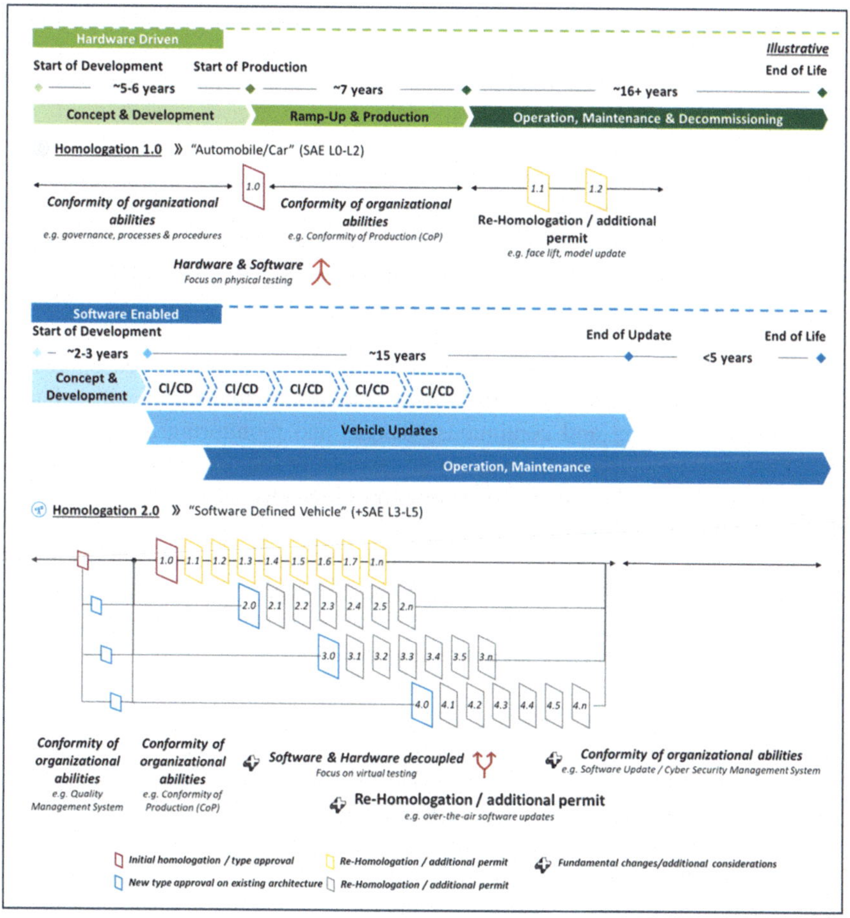

Fig. 5.1 Key elements of homologation/type approval over a vehicle's lifecycle. *Source* Own design

1. Concept and development (5 to 6 years): During this phase, the vehicle's core systems are designed, and regulatory requirements are first assessed. Initial software development and safety analysis occur here, setting the foundation for later stages.
2. Ramp-up and production (approximately 7 years): As the vehicle moves into production, homologation ensures that it meets all required standards, including testing and certification. Software updates become part of the production process, and any safety–critical software must be rigorously tested to meet regulatory standards.
3. Operation, maintenance, and decommissioning (16 + years): This phase involves the vehicle's entire lifecycle on the road. With SDVs, homologation continues as updates are delivered to vehicles via OTA, and safety–critical software must remain compliant throughout the vehicle's lifespan. As vehicles reach the end of their lifecycle, they must undergo decommissioning procedures that ensure they no longer pose safety risks.

In effect, homologation activities are shifting from hardware defined vehicles – with ~ 100 separate electronic control units, wherein all components are subject to comprehensive regulations and clear guidelines established to secure vehicle safety, energy efficiency as well as environmental norms—to SDVs. The latter are spanning autonomous driving level 3 to level 5 with centralized high performing computers and facing changing and sometimes unclear guidelines and even regulatory gaps related to data storage systems, software update procedures or cybersecurity norms. With this shift, virtual testing and simulation will enhance current physical testing and continuous changes and monitoring after a vehicle entered the market will become ordinary in contrast to today's selective even limited changes after market entry. To secure and efficiently verify and validate the vehicle over the entire lifecycle organizations will need to extend their scope coverage ensuring re-homologation and additional permits if new software packages and vehicle features come impact the previous type approval.

The shift to the lifecycle-wide view of homologation is essential to ensure SDV safety and legal conformity in an era of continual innovation. The vehicle's software must be continuously certified ("re-homologation" in Fig. 5.1) and updated to ensure compliance with functional safety standards, particularly as new features and functionalities, driven by AI and continuous software updates, are introduced. This evolving approach to homologation is critical for ensuring the ongoing safety of SDVs and aligns with broader industry trends toward more dynamic, flexible certification systems that can respond to rapid advancements in technology.

While in the hardware-driven homologation 1.0 process software and hardware having a strong focus on physical testing and rather looking at a linear subsequent process following the initial type approval with model updates which might require re-homologation, in the case of SDVs autonomous driving level 3 to 5 the concept and development phase shortens up and software capabilities will be decoupled from the E/E platforms being updated in shorter cycles, before the vehicle model will be phased out. With CI/CD digital homologation will become essential in a safety-focused environment as the one of automotive.

5.2.3 Requirements for the Technical Architecture in SDV Ecosystems to Guarantee Permanent Digital Homologation

To ensure digital homologation, the technical architecture in SDV ecosystems must fulfill various requirements. These include above all [1] a modular and scalable architecture and [2] open collaborations with defacto standards and open-source communities.

1. Modular and Scalable Architecture

While engineering excellence remains a core capability, it increasingly clashes with reusability and long-term efficiency. Many OEMs still operate under a "task force" mindset, where engineers work intensely to meet hardware-driven project deadlines, often delivering functionality that is hardcoded into ECUs. While these efforts may be technologically impressive, they often result in rigid systems that cannot be reused across vehicle generations. This architectural rigidity stems from legacy systems and fragmented development approaches, where functionally is deeply embedded in individual ECUs. As a result, adapting or upgrading software for subsequent vehicle models becomes time- and resource-intensive.

SDVs therefore utilize hypervisors and virtualization techniques to effectively separate safety-relevant functions from non-relevant ones, ensuring robust system architecture and operational safety. Technologies like QNX, AGL, and Xen are employed to create isolated environments where relevant/critical applications can run independently of non-critical ones, thereby enhancing system reliability and security. Hypervisors act as a thin layer of operational software that provide the supporting infrastructure for virtualization, allowing virtual machines to communicate with host hardware and connected peripherals. AGL and Xen enable the creation of standardized, hardware-agnostic environments and support the development of modular, safety-certified software. Xen provides the foundation for partitioning workloads, while AGL fosters community-driven, open-source innovation for automotive applications.

The shift toward high-performance computing is also reshaping traditional supplier relationships (see Chap. 4 in this book). Previously, OEMs sourced multiple ECUs from Tier 1 suppliers. However, with centralized computing, major semiconductor manufacturers are playing a more prominent role. Instead of integrating numerous ECUs, OEMs are shifting toward domain-based, zonal, or centralized architectures (see Sect. 4.3 in this book), enabled by powerful central computers. This transition requires clear architectural alignment and a redefinition of supplier roles.

This transformation also has substantial business and procurement implications. The shift from a traditional supply chain, OEM to Tier 1 to Tier 2, to a more dynamic (innovation) ecosystem with independent partners (see Chap. 2 in this book) requires new purchasing strategies. Instead of managing a broad network of smaller suppliers, OEMs must now negotiate directly with major semiconductor companies, which operate with different pricing structures, lead times, and risks. As a result, procurement strategies must adapt to secure long-term ecosystem partnerships, manage cost structures differently, and address potential supply bottlenecks in the semiconductor market.

Containers as Docker and Podman as packages of software containing all necessary elements to run in all environments, are adopted for deploying non-critical software, enabling faster development cycles and more efficient resource utilization. Containerization allows developers to package applications with their dependencies into portable containers, ensuring seamless operation across different environments and eliminating compatibility issues [5]. This approach facilitates rapid scalability and deployment, particularly beneficial for applications experiencing fluctuating loads, such as e-commerce platforms during high-traffic events.

The SOA approach (see Chap. 4) allows for microservices that facilitate independent development, deployment, and scalability, enhancing the flexibility and efficiency of SDV innovation ecosystems. Microservices architecture breaks down large systems into smaller, independent services, allowing teams to develop and deploy services separately, ensuring that issues in one component do not affect others. This modular design enhances scalability and resilience, enabling organizations to respond swiftly to market changes or project demands.

Greenfield OEMs, unburdened by legacy constraints, demonstrate an alternative approach. By adopting centralized architectures and reducing the number of ECUs, by installing few HPCs, they streamline development, enable feature reusability, and drastically reduce system complexity. According to the Deloitte Readiness Study 2024 [6], this simplification offers significant economic advantages, as shown in Fig. 5.2. OEMs that operate with fewer than five software versions and implement streamlined architecture report up to 36% lower SDV-related R&D expenditures compared to those maintaining more than ten versions. These cost savings stem from:

- Reduced integration and testing overhead,
- Easier system maintenance and OTA update capability,
- Shared and reusable software modules across platforms and models, and
- A shift toward modular, containerized development approaches that facilitate parallel workflows and reduce bottlenecks.

These findings highlight that streamlined and modular software architectures especially when paired with reduced software versioning and containerization, are essential technical prerequisites for ensuring permanent digital homologation with scalable SDV ecosystems.

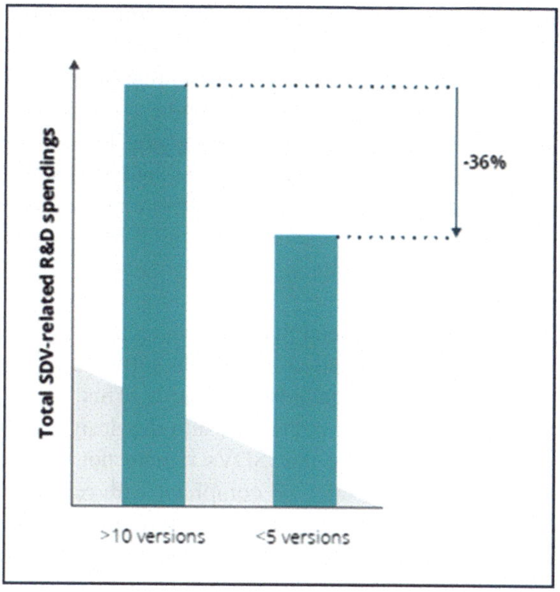

Fig. 5.2 Impact of software version reduction on SDV-related R & D expenditure. *Source* Deloitte, 2024 [2]

2. Open collaborations with DeFacto Standards and Open-Source Communities

In SDVs, continuous compliance cannot be achieved solely through internal development. Open-source communities and standards have become crucial enablers in the development of SDVs, as they foster collaboration and innovation across the automotive ecosystem. The nature of SDVs requires the integration of a vast range of software and hardware systems. Open-source platforms play a pivotal role in driving these advancements, particularly by creating standardized frameworks, tools, and protocols. These platforms not only enhance the functionality and performance of SDVs but also accelerate their adoption by allowing different players in the ecosystem to contribute and build upon existing solutions.

One of the key initiatives is the Eclipse SDV Working Groups, a working group within the Eclipse Foundation, which focus on the development of open-source platforms, middleware, and tools specifically designed for SDVs. These working groups facilitate collaboration among automotive manufacturers, technology providers, and open-source communities to create flexible and scalable solutions that support the complex needs of SDVs [7]. Through these platforms, stakeholders can share resources, tools, and best practices, reducing duplication of efforts and improving the speed of innovation. Furthermore, middleware and tools developed within these working groups enable more seamless integration between various software layers in SDVs, ensuring that disparate systems can work together effectively and efficiently.

In the highly interconnected and dynamic world of SDVs, digital homologation has become an essential requirement. The ability to make continuous updates and adjustments to vehicle software presents significant challenges in terms of safety, interoperability, and compliance. Therefore, a well-thought-out technical architecture is indispensable to ensure that new software functionalities can be introduced without compromising the safety and regulatory compliance of the vehicles.

Such an architecture must provide a robust infrastructure for OTA updates, ensuring that new software versions can be distributed securely and efficiently. Additionally, standardized interfaces and protocols are necessary to enable the seamless integration of different software and hardware components. This requires a high level of collaboration between OEMs, suppliers, and open-source communities to develop common standards that can serve as the foundation for digital homologation.

Moreover, it is crucial that the architecture includes mechanisms for continuous monitoring and validation of vehicle software to ensure that all systems comply with the required regulatory standards. This means that safety and compliance requirements must be embedded into the design phase of new software solutions. Such measures ensure that SDVs remain not only technologically advanced and innovative but also safe and compliant with regulations.

In addition to the Eclipse Working Groups, AUTOSAR (AUTomotive Open System ARchitecture) Adaptive and SOAFEE (Scalable Open Architecture for Embedded Edge) are key standards that provide standardized interfaces for automotive software, making it easier to develop and integrate software applications within SDVs. AUTOSAR Adaptive, an evolution of the classic AUTOSAR standard, focuses on enabling software-defined features such as autonomous driving, ADAS, and infotainment. It offers a platform for high-performance computing in SDVs, allowing developers to create software solutions that can operate across different hardware platforms while ensuring safety, scalability, and compliance with automotive standards. SOAFEE, on the other hand, aims to provide a scalable and cloud-native architecture for edge computing in vehicles. By offering a standardized framework for deploying software in an automotive environment, SOAFEE supports the integration of advanced technologies like AI, machine learning, and real-time data processing.

To foster the growth of the SDV ecosystem, there is a need for hardware-agnostic, standardized software interfaces. These interfaces enable scalability and allow the same application to be deployed across different makes and models. To encourage third-party developers to contribute, an automotive development suite must be established. The creation of such a development and emulation suite, which abstracts the underlying hardware, was a key success factor in the smartphone app ecosystem. Implementing this approach in the automotive sector will accelerate innovation and improve interoperability. Notably, digital.auto is an open community that promotes a product- and vendor-neutral initiative.

Another prominent example is RHIVOS (Red Hat In-Vehicle OS), which is focused on containerization and cloud-native approaches for SDVs. Containerization enables the deployment of software in lightweight, isolated environments, allowing developers to package and run software in a consistent manner across different platforms [8]. This is particularly important for SDVs, as it enables more efficient management of software updates, enhances security by isolating components, and facilitates the integration of new functionalities. RHIVOS provides an operating system that can handle both in-vehicle and cloud-based workloads, ensuring that SDVs can operate seamlessly in a hybrid environment [9]. This architecture also makes it easier to manage the complex software stacks required for autonomous driving, real-time analytics, and other critical SDV applications.

Lastly, the integration of middleware and abstraction layers is crucial for ensuring interoperability between the diverse hardware and software systems within SDVs ecosystem. These systems act as communication bridges between software components, hardware modules, and external services, facilitating their seamless operation without conflicts. By abstracting the complexities of underlying hardware, these layers enable software applications to function without requiring significant modifications. This is particularly important in SDVs, where interactions between sensors, control systems, AI models, and cloud services must be reliable and efficient. With this approach, vehicles can maintain consistent performance across different models, manufacturers, and technology providers, even as the underlying microprocessor and microcontroller-based control units evolve to handle increasingly complex workloads.

In effect, the SDV ecosystem is becoming more accessible through open-source initiatives and standardization efforts (see Fig. 5.3). The percentages listed (e.g., for AGL, COVESA) reflect the share of surveyed organizations reportedly using each framework or platform as part of their SDV software stack, based on internal industry study data as examined in the Global SDV survey representing the feedback of over 160 OEM executives [6].

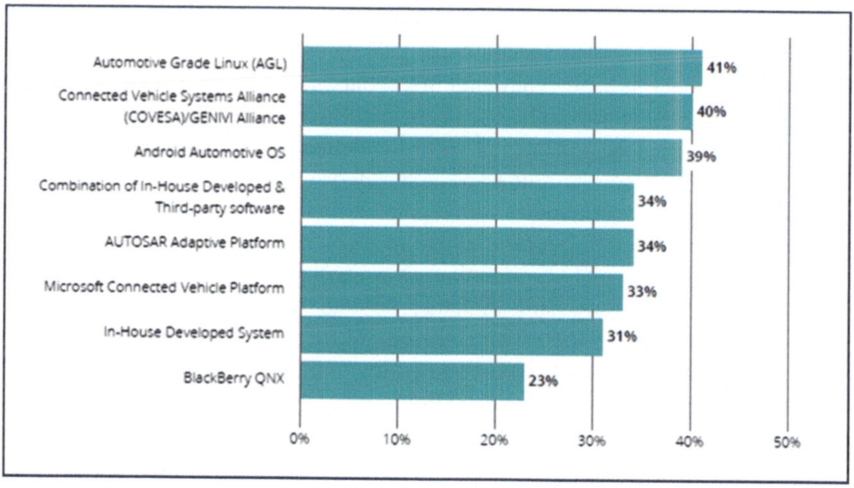

Fig. 5.3 Use of frameworks/global operating systems for SDV implementation (in %) *Source* Own design

Organizations like the Automotive Grade Linux (AGL) (41%), Connected Vehicle Systems Alliance (COVESA/GENIVI) (40%), and Android Automotive OS (39%) are leading the way in creating modular, reusable components that simplify hardware complexity and enable a flexible software ecosystem. Other platforms such as the AUTOSAR Adaptive Platform (34%), Microsoft Connected Vehicle Platform (33%), and BlackBerry QNX (23%) also play a crucial role in this evolving landscape. Moreover, the combination of in-house developed and third-party software (34%) and fully in-house developed systems (31%) highlights the diverse approaches taken by OEMs and suppliers. Low-code/no-code development platforms are also emerging, making SDV development accessible to non-automotive software developers, thereby lowering the entry barrier for innovation in the automotive industry.

The automotive industry is at a critical juncture in its transition to SDVs. While the potential for innovation is immense, the above-mentioned requirements for the digital architecture in SDV ecosystems must be addressed to guarantee digital homologation.

While the industry should focus on standardizing the non-differentiating aspects of SDVs, regulatory bodies will still play a key role. In particular, safety and cybersecurity regulations must evolve to address the new challenges posed by SDVs. OTA updates especially for the current ADAS content for example, will need to comply with both safety standards and cybersecurity protocols. This is crucial for ensuring that vehicles remain secure and functional throughout their entire lifecycle, including several years after initial production.

A critical gap in current regulations is the lack of a comprehensive framework for ADAS in the context of OTA updates. As ADAS features become more advanced and rely increasingly on software updates for performance improvements and bug fixes, regulatory oversight must ensure that these updates do not compromise safety. The industry needs clear guidelines on how changes to ADAS functionalities, such as emergency braking or lane-keeping assistance, should be tested, validated, and approved before being deployed OTA. Without such a framework, there is a risk that incremental software changes could unintentionally alter system behavior in ways that impact vehicle safety.

Hence, OEMs must collaborate with regulatory bodies to ensure that updates are not just feature-enhancing, but also adhere to the necessary safety and cybersecurity requirements. This includes defining clear regulations for the communication between onboard and offboard systems, particularly in the context of OTA updates. Standardized protocols must ensure secure and reliable data transmission between the vehicle and backend infrastructure, minimizing risks associated with unauthorized access or data corruption.

5.2.4 Differentiating SDV Architecture Aspects Compared to Smartphone and Computer Ecosystems

Despite its complexity, SDVs offer something unprecedented in automotive history: the ability to modify and enhance the product post-sale (see Chap. 4 in this book). For the first time, vehicles can be upgraded and redefined through software updates, allowing manufacturers to adapt to changing needs and technologies. But to fully capitalize on this, the complexity must be managed effectively, ensuring that OEMs can meet the escalating demands for innovation without compromising safety, performance, or cost-efficiency.

This is where the difference between SDVs and consumer electronics becomes clear. While smartphones and computers also experience rapid evolution through software, SDVs face unique challenges that stem from their inherent need for safety, real-time processing, and specialized system architectures. These differences are crucial in shaping how automotive systems are designed, built, and maintained. One of the most critical distinctions is the necessity for real-time processing and deterministic behavior in SDVs. In contrast to smartphones and computers, where latency variations are generally tolerable, SDVs must meet stringent real-time constraints, particularly for functions directly impacting safety and performance.

While consumer electronics primarily rely on multitasking operating systems that prioritize user experience, SDVs require hard real-time computing to guarantee the timely execution of safety–critical tasks. The need for deterministic behavior ensures that the system responds within a known time frame, making it predictable and reliable in situations where lives are at stake [10].

A key difference is in technical architecture. Consumer devices, such as smartphones and computers, generally utilize a centralized architecture with a few powerful processors. In contrast, SDVs are transitioning from distributed ECUs to more advanced zonal architectures and centralized computing systems. This shift enables more efficient data processing by centralizing various vehicle control tasks that were traditionally handled by individual ECUs. The move to centralized systems allows for better integration, improved performance, and simpler maintenance.

SDVs also blend edge processing with cloud connectivity to facilitate real-time decision-making and continuous updates. While consumer electronics mainly rely on local processing, SDVs leverage both onboard processing for immediate decisions and cloud-based services for continuous OTA updates and fleet management. This combination enables real-time monitoring, diagnostics, and software updates, which are essential for optimizing vehicle performance and maintaining up-to-date features.

Finally, security and safety in automotive systems are governed by stringent and industry-specific standards, most notably ISO/SAE 21,434 for cybersecurity and ISO 26262 for functional safety, which go far beyond the practices seen in consumer electronics. These standards ensure the integrity of SDV systems, mitigating risks associated with software vulnerabilities and hardware failures. While consumer electronics also prioritize security, the automotive industry must meet more stringent requirements due to the high risks associated with system failures in transportation [11].

The differences of the SDV architecture compared to smartphone and computer ecosystems thus follow from the challenges of digital homologation for the design of technical architectures in SDV ecosystems, which were outlined in this chapter. In addition to these technical considerations, collaboration must be worthwhile not only in innovation ecosystems such as for SDV, but also in transaction ecosystems. In the next two chapters, starting points for joint value creation with seamless intermodal mobility offers (in MaaS transaction ecosystems) are shown, because these have not yet been profitable.

5.2.5 Opportunities for Business Building in Transition to SDV

As the automotive industry transitions toward SDVs, the focus is not limited to technical improvements such as system stability, security, and OTA update capability. It also includes the creation of entirely new business models. A key enabler in this transformation is the establishment of an in-vehicle App Store or digital marketplace, allowing OEMs and partners to offer software-based services and features that can be downloaded, activated, or subscribed to directly from within the vehicle.

This shift creates a strategic foundation for generating business beyond the initial vehicle sale. OEMs can monetize software over the entire lifecycle of the vehicle by introducing models such as function-on-demand models for features like heated seats, enhanced navigation, or ADAS. They can also offer subscription-based services for entertainment, traffic information, or security, along with third-party app ecosystems that resemble established platforms such as the Apple App Store or Google Play. This opens the door for collaboration with external developers, startups, and content providers who can build value-added applications tailored to user preferences, regional demands, or fleet needs. Ultimately, the in-car digital marketplace transforms the vehicle into a living, upgradable platform, aligning with user expectations shaped by the smartphone era, and positioning automotive players to participate in the broader MaaS and digital services economy.

Early attempts by OEMs to monetize SDVs through in-vehicle app stores have largely failed to create meaningful differentiation. Most current marketplaces are limited to infotainment services such as navigation, streaming, or news, offering little that a smartphone can't already deliver more seamlessly.

As a result, today's digital marketplaces are peripheral at best. They do not tap into the unique value of the vehicle itself, nor do they offer a compelling alternative to dominant tech ecosystems. Up to now, there is no data advantage, and no strategic control over user experience or software innovation.

One of the most powerful opportunities lies in vehicle-level customization. Imagine a driver selecting a Formula 1-style dashboard theme or activating an enhanced lane-keeping system developed by a certified third-party provider. Today, such use cases are blocked by rigid homologation and closed architectures. But with proper abstraction, modular certification, and OTA capability, these barriers can be overcome.

In this vision, the vehicle itself becomes the platform, not just a container for entertainment apps, but a programmable, evolving system that adapts to drivers' preferences, driving conditions, and use cases.

5.2.6 The Possible Future Role of Traditional Automotive Companies in Transition to SDV

The transition to SDVs challenges traditional automotive companies (OEMs) to go beyond their legacy role as product manufacturers and assume a far more strategic and systemic function: That of platform orchestrators and guardians of vehicle functionality. In this role, traditional OEMs do not merely produce connected hardware. Instead, they curate secure ecosystems, ensure regulatory compliance, enable innovation, and abstract complexity for developers and end users alike.

To achieve this, traditional OEMs must embrace a dual central role as orchestrators and guardians in SDV innovation ecosystems, both from technical and strategical perspectives. Rather than focusing narrowly on infotainment, traditional OEMs must face three challenges: These OEMs first have to curate a vehicle-wide marketplace that goes far beyond media and apps, extending into the car's core functionalities. Secondly, they have to enable third-party development not only at the HMI layer, but also for ADAS, domain-specific control functions, and personalized cockpit experiences. Thirdly, abstract safety, compliance, and security in a robust platform layer have to ensure that third-party innovation does not compromise functional safety or homologation. This orchestration role requires OEMs to move from passive gatekeepers to active enablers, offering SDKs, APIs, and controlled access points, while shielding sensitive vehicle layers through strict governance, certification mechanisms, and real-time monitoring.

Realizing this orchestration role requires both, technical and regulatory innovation. On the one hand, looking at the technical foundations of modular, service-oriented architectures, secures vehicle operating systems and hypervisors as well as policy-driven access control and sandboxed third-party execution. On the other hand, establishing the legal and compliance frameworks will redefine homologation to allow for dynamic, software-based updates, establish certification mechanisms for third-party vehicle function providers and ensure cybersecurity, data privacy, and liability management. By assuming responsibility for these foundational elements, OEMs position themselves not as mere integrators but as trusted orchestrators of the mobility ecosystem.

Traditional OEMs need to also cover a key dimension of this role: Guardianship. Unlike app stores in consumer electronics, where user experience is paramount, the vehicle environment must prioritize safety, reliability, and resilience. OEMs must vet software updates and third-party components rigorously, enforce real-time monitoring and rollback mechanisms and protect vehicle integrity and ensure cross-domain system stability. This guardianship extends to data sovereignty, functional separation, and even ethical design. In these areas, traditional OEMs, not Big Tech, should take the lead.

Note

1. Differentiation of Safety Levels (e.g. ASIL B versus ASIL D): Automotive Safety Integrity Levels (ASIL) categorize the risk associated with vehicle functions. ASIL B represents a moderate risk level, applicable to infotainment and non-critical ADAS functions. ASIL D, on the other hand, is the highest safety level, required for critical systems like braking, steering, and automated driving. ASIL-D systems demand redundancy, real-time operating systems (RTOS), and certified software development processes, whereas ASIL-B functions may utilize Linux-based or commercial operating systems with less stringent requirements.

References

1. Xian, X., & Chian-Ku, F. (2019). *Autonomous vehicles, risk perceptions and insurance demand: An individual survey in China.* https://doi.org/10.1016/j.tra.2018.04.009
2. Deloitte. (2023). *More autonomous trucks are hitting the road: How should insurers be changing lanes on coverage?* Retrieved May 2, 2025, from https://www.deloitte.com/us/en/insights/industry/financial-services/financial-services-industry-predictions/2023/fsi-autonomous-vehicle-insurance-predictions.html
3. King, C. (2022). *Bewertung von Fahrerassistenzsystemen im Umfeld des szenariobasierten Testens.* https://doi.org/10.5445/IR/1000157665
4. Dengle, A. (2024). *Driving the future: Transforming homologation and regulatory compliance.* ISG. Retrieved May 2, 2025, from https://www.nttdata.com/global/en/-/media/nttdataglobal/1_files/industries/automotive/isg-white-paper-driving-the-future-2024-03-06-v1.pdf?rev=c7e649041d484ae09d91118ec371c33b
5. Cherukuri, B. R. (2024). *Containerization in Cloud Computing Comparing Docker and Kubernetes for Scalable Web Applications.* https://doi.org/10.30574/ijsra.2024.13.1.2035
6. Deloitte. (2024). *Global manufacturer readiness study SDV.* Retrieved May 2, 2025, from https://www.deloitte.com/et/en/Industries/automotive/analysis/software-defined-vehicles.html
7. The Eclipse Foundation. (2025). *Software Defined Vehicle – about the working group.* Retrieved May 2, 2025, from https://sdv.eclipse.org/
8. Marzani, S., Chunduru, P., & Mager, M. (2024). *Executing container orchestration with Eclipse BlueChi on the RedHat in-vehicle operating system in the AWS cloud.* AWS for Industries. Retrieved May 2, 2025, from https://aws.amazon.com/blogs/industries/executing-container-orchestration-with-eclipse-bluechi-on-the-redhat-in-vehicle-operating-system-in-the-aws-cloud-part-1
9. Chow, F. (2022). *The new standard: Red Hat In-Vehicle Operating System in modern and future vehicles.* Red Hat. Retrieved May 2, 2025, from https://www.redhat.com/en/blog/new-standard-red-hat-vehicle-operating-system-modern-and-future-vehicles

10. Kumar, S. V. G., Vadivel, M., Penchalaiah, U., & Ganesan, P. (2019). *Real time embedded system for automobile automation.* IEEE. https://doi.org/10.1109/ICSCAN.2019.8878820
11. De Vincenzi, M., Pesé, M. D., Bodei, C., Matteucci, I., Brooks, R. R., Hasan, M., Saracino, A., Hamad, M., & Steinhorst, S. (2024). *Contextualizing security and privacy of software-defined vehicles: State of the art and industry perspectives.* https://doi.org/10.48550/arXiv.2411.10612

Mobility as a Service Transaction Ecosystems Worldwide

6

Arne Jeppe, Arman Arzani, Marcus Handte, Nico Zauner, Pedro J. Marrón, Fabian Langanki, and Heike Proff

Contents

A. Jeppe (✉) · A. Arzani · M. Handte · P. J. Marrón · F. Langanki · H. Proff
University of Duisburg-Essen, Duisburg/Essen, Germany
e-mail: arne.jeppe@uni-due.de

A. Arzani
e-mail: arman.arzani@uni-due.de

M. Handte
e-mail: marcus.handte@uni-due.de

P. J. Marrón
e-mail: pjmarron@uni-due.de

F. Langanki
e-mail: fabian.langanki@uni-due.de

H. Proff
e-mail: heike.proff@uni-due.de

N. Zauner
Deloitte, Düsseldorf, Germany

IESE Business School, Barcelona, Spain

H. Proff et al. (eds.), *Future of the Automotive Industry*, Future of Business and Finance, https://doi.org/10.1007/978-3-032-26308-7_6

Abstract

The transformation in the automotive industry from system integration to ecosystem play is not only being driven by increasing competition from technology companies in the field of Software-defined Vehicles (SDV) via platforms. There is also a transformation to shared mobility, driven by car and ridesharing companies. They often offer Mobility as a Service (MaaS) as a seamless offering of different mobility services, e.g., public transport, car sharing, bike sharing, taxi. MaaS is desired – worldwide, albeit with regional differences – by customers and transport planners, because it is essential for achieving the net-zero goals and helps to reduce road traffic especially in city centers. This chapter uses Natural Language Processing (NLP), an AI-based text analysis method, to examine trends in the development of MaaS ecosystems between 2015 and 2023 in order to understand why MaaS has not et been able to be offered profitably.

6.1 Introduction

As mentioned in Chap. 1, the transformation in the automotive industry from system integration to ecosystem play is not only being driven by increasing competition from technology companies in the field of Software-defined Vehicles (SDV) via platforms (see Chaps. 4 and 5). There is also a transformation to shared mobility, driven by car and ridesharing companies. They often offer Mobility as a Service (MaaS), which we have already defined as a seamless offering of different mobility services [25], e.g., public transport, car sharing, bike sharing, taxi. MaaS is desired – worldwide, albeit with regional differences – by customers [23] and transport planners, because it is essential for achieving the net-zero goals [43] and helps to reduce road traffic especially in city centers [3].

MaaS that not only integrates different mobility services from different providers, but also "provides one-stop access to these services through a common interface" ([32], p. 283) is in line with social trends such as digitalization, service orientation, and the sharing economy. It also corresponds to the first three key components of ecosystems according to Adner [1]: (1) an overarching value proposition (that goes beyond the value propositions of the partners), (2) an operating model of partners and relationships and (3) a technical platform (see Chap. 2). More precisely, MaaS needs a transaction platform in a transaction ecosystem [38], i.e., a multi-sided marketplace enabling the reuse of "common resources" (activities and assets) of the ecosystem partners [12].

These MaaS transaction ecosystems are not (yet) profitable and in practice worldwide and most of the many pilot projects do not reach the implementation stage [32]. In March 2024, even the Finnish mobility startup MaaS Global, founded in 2015 as the first provider of a MaaS app (Whim) had to file for insolvency after Covid and the financial crisis [40]. That is why we need more insights into MaaS ecosystems and their problems [6] in order to understand their opportunities for joint value creation (see Chap. 2 in this book).

This chapter analyzes the most important MaaS transaction ecosystems worldwide by identifying patterns in their development since 2015 [29], one year after their first introduction by Sampo Hietanen (2014) [25]. However, this requires an innovative methodology (see endnote 1), as it is difficult to analyze MaaS ecosystems with many, often small mobility service providers such as bicycle, ridesharing or small car-sharing providers [31]. This is because these providers are usually not subject to the publication obligation [7] or publish only limited [20] and often only unstructured information [21].

We, therefore, use an AI-based text analysis, i.e., a NLP method that is already used in management sciences [11, 10] and that has been innovatively extended by a NLI classification that requires little training data. This method (see endnote 1) makes it possible to use machine learning algorithms to recognize patterns in limited unstructured information classification for training over several years so that changes can be analyzed. As an "uncovering method," NLP can be seen as a first step toward practical improvement and in-depth theory building [42]. We analyze 86,000 news articles from 200 countries stemming from the Factiva database [9], supplemented by annual and analyst reports from the MaaS ecosystem.

In this chapter, we first introduce the NLI-based classification method that we propose as an innovative NLP application for the study of MaaS ecosystems (Sect. 6.2). Then we explain how we implemented this new method to study MaaS ecosystems (Sect. 6.3): First by defining indicators for MaaS ecosystems, then by collecting, filtering, classifying, and validating data. Finally, we summarize the identified patterns of important MaaS ecosystems worldwide and their development since 2015 as the result of data analysis and interpretation (Sect. 6.4) and discuss implications (hints for improving MaaS ecosystems and its indicators as well as recommendations for more structural changes toward a successful design of transaction ecosystems for MaaS; Sect. 6.5).

6.2 New Application of the NLP Method to study MaaS Ecosystems

As mentioned already, it is not easy to collect information about activities in ecosystems with often small partner companies [13], when only limited and unstructured information can be found. Case studies can only capture a small section of reality [42] and cannot avoid a certain subjectivity. Traditional content analyses of texts or documents, in which a coder examines categories, also remain subjective [36].

Therefore, advanced text analyses based on machine learning/AI found their way into management [42]. These methods allow the software to recognize patterns based on powerful algorithms, even if specialized humans are still needed to guide study design and interpretation [10]. Above all, NLP is gradually being taken up in management science [11, 10, 42]. It enables the detection of patterns in MaaS ecosystems in six steps (See Fig. 6.1):

step (1) indicator definition, then step (2) data acquisition, step (3) filtering, step (4) classification and step (5) validation of data as well as step (6) data analysis and interpretation (see Fig. 6.1). Steps (3) to (5) are each carried out by an indicator iteration with the sub-steps of constructing training data (steps 3a, 4a, and 5a), developing possible classifiers (steps 3b, 4b, and 5b) and comparing the classification models (steps 3c, 4c, and 5c), see also Fig. 6.1.

Since many NLP applications are less suitable for limited, unstructured information [21], as is the case for MaaS ecosystems, we used a NLI [30] application that enables accurate NLI-based classification without requiring large datasets for training (see endnote 1). It offers the advantages that input and inference can be individualized and a high abstraction capacity with low data availability and high generalizability is possible [15].

Extensive training data for the automation of NLI could be avoided by an NLI-based classification based on Large Language Models (LLMs [45]), which is already available in computer science and has occasionally emerged as a method in management research. This method is based on deep neural networks [4] and can capture the statistical properties of many languages (see endnote 1). An NLI-based classification based on LLMs enables an interplay of AI-based learning and joint training and verification by mobility strategy researchers and computer scientists, which is also possible with little unstructured data [22].[1]

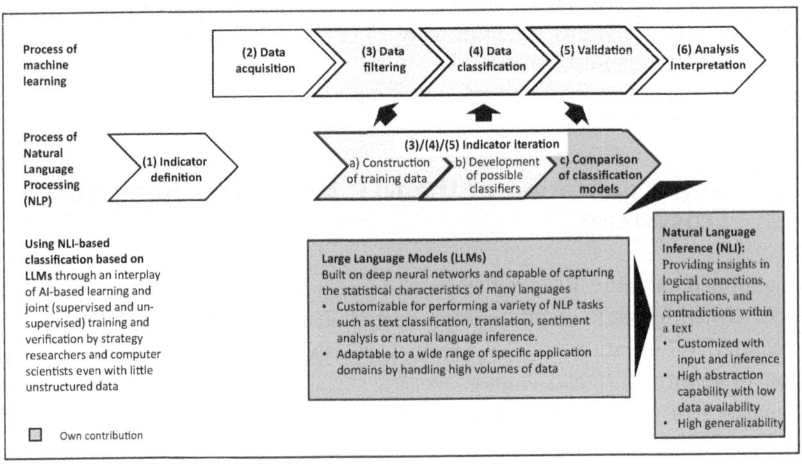

Fig. 6.1 Using NLI-based classification for NLP in management. *Source* See also [21]

6.3 Applying the Innovative NLP Application to investigate MaaS Ecosystems

To recognize patterns in MaaS ecosystems worldwide, we performed an NLI-based classification along the six steps in Fig. 6.1 (see also endnote 1): First the definition of indicators for MaaS ecosystems (Sect. 6.3.1) then the acquisition, filtering, classification, and validation of data (Sect. 6.3.2). The results of the data analysis and their interpretation are presented in Sect. 6.4.

6.3.1 Definition of Indicators of MaaS Ecosystems

Defining indicators of MaaS ecosystems means defining reference points for the semantic-based analysis. They can be derived based on the key constructs of (transaction) ecosystems: First, we used the constructs of platform-based ecosystems according to Adner [1], which are also included in the definition of special MaaS ecosystems [31], see Chap. 2 in this book: (I) an overarching value proposition, (II) an operating model of the partners with data and financial flows between them and (III) the technical platform. In addition, we took constructs from the explanations of joint value creation in ecosystems ([25, 28, 16] and see Fig. 2.4 in Chap. 2): (IV) governance mechanisms and (V) value drivers.

To record ecosystems, we divided these five key constructs into a total of 22 variables (see Table 6.1 based on Fig. 2.4 in Chap. 2), based on the approaches offered to explain the individual categories in Chap. 2 (Fig. 2.4, see also endnote 1):

- The value proposition construct was divided into three variables integrated, individualized and interactive customer solutions (as the basic benefit) in accordance with the service-dominant logic [44] and an additional benefit (possible e.g., through – data protection – compliant-use of the data or advertising).
- The operating model was divided into the six variable partners (orchestrator/complementors, public/private), relationships, asset investments, asset collaboration, financial flows, and data flows [34].
- The construct platform is formed from the two variables platform design (interface for interaction, back and forth) and functionality (e.g., payment function) [19].
- Four governance mechanisms were differentiated according to the relational view [16] (see also Sect. 2.1): Effective governance through rules, regulations and roles of partners; investments in interfirm relation-specific assets, above all trust and power as well as interfirm knowledge-sharing routines as well as the variable governing partner.
- The variables of the value driver construct are derived from the six platform effects ([24], see also Sect. 2.3): modularization [28] in accordance with transaction cost theory; complementary resources and capabilities, in accordance with the relational view [16]; value co-creation with customers in accordance with the service-dominant logic [44], scaling the network in accordance with network theory [33], cross-subsidizing the more price-sensitive side in accordance with the theory of multi-sided markets [17], and AI-based learning [27] in accordance with explanations of learning.

Table 6.1 Constructs and variables used in the study. *Source* Based on endnote 1 and Fig. 2.4 in Chap. 2 and Fig. 2.4 in Chap. 2

Constructs	(I) Focal value proposition	(II) Operating model	(III) Technical platform	(IV) (Internal) Governance mechanisms	(V) Value drivers
Variables	Individual Customer Solution	Partner – Orchestrator/ Complementors – Public/Private	Platform Device (Technology) – Interface for Interaction – Back-end Systems	Effective governance – Rules – Regulations – Roles of Partners	Modularization Complementarity
	Integrated Customer Solution	Relationships	Functionality of the Platform – e.g. Payment Func.	Inter-firm relation-specific assets – Trust – Power	Value co-creation with customers Scaling the Network
	Interactive Customer Solution	Asset Investment		Inter-firm knowledge-sharing routines	Cross-Subsidization of weak Partners in a Network
	Additional Benefit	Asset Collaboration		Governing the set of Partners	AI-based Learning
		Financial flows			
		Data flows			

6.3.2 Acquisition, Filtering, Classification, Validation of Data

A systematic search with five sub-steps was first carried out to collect and filter data on MaaS ecosystems worldwide (Fig. 6.2) before these could then be classified and validated. The Factiva database (see endnote 2) with 672 million news articles from various news sources worldwide was an obvious choice. We supplemented annual and analyst reports of MaaS ecosystems that exist in the MaaS industry to reduce bias from external media coverage [18]. We examined the period from 2015 to 2023 because the MaaS concept was only introduced in 2014 by the CEO of the Intelligent Transportation Society of Finland, ITS (Hietanen, 2014, [25]) and 2023 was the last year for which complete data was available. We used English and German sources because we could only validate these languages.

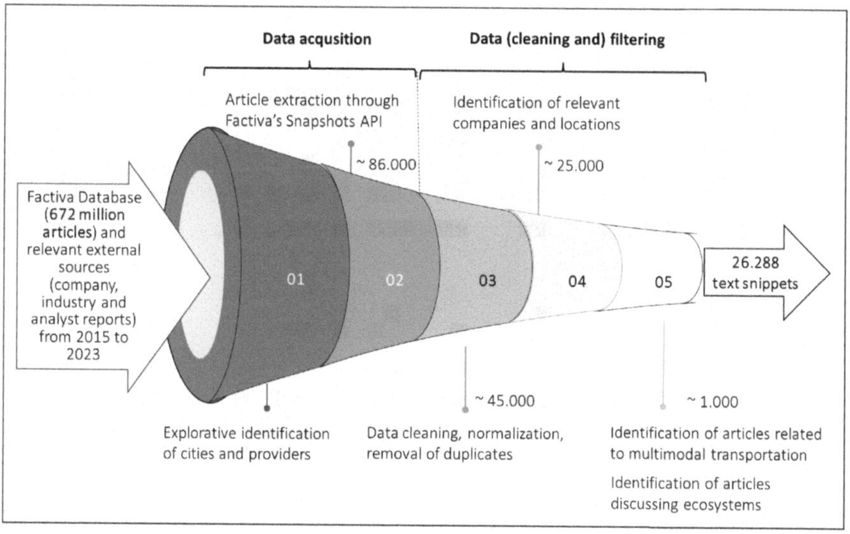

Fig. 6.2 Process of data acquisition and data filtering MaaS ecosystems worldwide. *Source* See endnote 1

Data acquisition (Step (2), Fig. 6.1): For data acquisition (also Step 01 in Fig. 6.2), we were able to identify a total of 28 MaaS providers and 40 cities worldwide with MaaS (Fig. 6.3) using a systematic exploratory search in the Factiva database for articles mentioning MaaS offerings (such as Berlin or Helsinki) and companies/ providers as coordinator or orchestrator [39] in these ecosystems (such as Jelbi, Trafi, or Whim).

City and provider groups were then used to estimate the potential volume of data for data extractions in step 02 (Fig. 6.2) to ensure data quality (see endnote 1). We also refined some of our queries using the company codes (unique company identifiers) provided by the Factiva application programming interface (API). Thus, we were able to extract 86,000 news articles in a Factiva query that mention at least one provider and one city for each city and provider from Fig. 6.3.

Data (Cleaning and) Filtering (Step (3), Fig. 6.1): Since text content is often unstructured, we performed data (cleaning and) filtering (also step 03 in Fig. 6.2), [41] and removed articles that consisted only of headlines or duplicate news articles that were either the result of updates or secondary publication. To do this, we converted the documents into a vector representation and applied in-exact similarity matching to the resulting embeddings [35]. After deduplication, 45,000 articles remained.

We then performed Named Entity Recognition (NER [35]) on step 04 in Fig. 6.2 to extract the names of the cities and providers mentioned in the documents. Finally, we removed all documents that did not mention at least one MaaS provider and one target city. In total, this data cleaning resulted in a list of about 25,000 unique news articles mentioning relevant providers and cities worldwide.

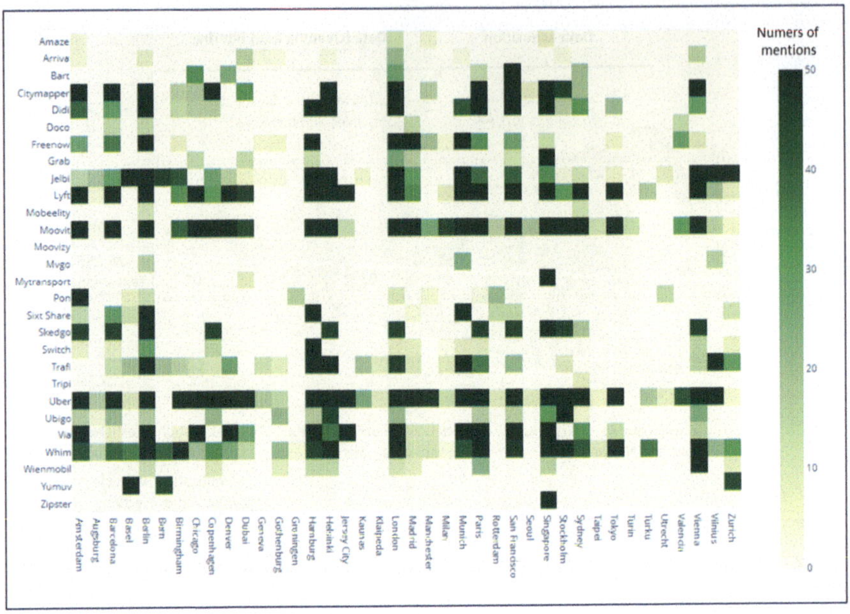

Fig. 6.3 MaaS providers and cities worldwide. *Source* Own design

In addition, all articles that did not refer to multimodal transportation and MaaS ecosystems were removed (step 05 in Fig. 6.2) by AI-based data filtering with an indicator iteration with an NLI-based classification.[2] The reason for introducing this additional filtering step is that a news article mentioning a specific provider does not necessarily indicate an article that reports on MaaS ecosystems as defined by Adner ([1]: Chap. 2). After that, only 1000 news articles remained. As the little information is distributed in a very unstructured way, we divided these articles into a total of 26,288 snippets and then labeled them.

Data Classification and Validation (steps (4) and (5) in Fig. 6.1):

To support the identification of constructs and their variables, a labeling and validation tool was developed. The classification was carried out between February 2023 and February 2024 by two programmers as part of a collaborative project between the authors of this chapter – business economists and computer scientists working in the field of NLP (see endnote 4). Since we are using pretrained LLMs that are fine-tuned for NLI using existing datasets, it is possible to apply our analysis to the MaaS domain, even without the existence of a large training dataset related to MaaS.

However, to ensure a high-quality analysis, it is important to assess and validate the outputs thoroughly.[3], [5] By manually validating 10% of the classification output for each variable, we calculated quality metrics such as the F1 score (a metric balancing precision and recall) [2]. To achieve high precision and recall, the pipeline was defined and repeatedly optimized.[4] As a result, the structure of the variables could be semantically more accurately captured, thereby improving the F1 scores (see endnote 1). The process was repeated until the experts were certain of the quality of the classifiers for each given construct. This enabled us to show that quality criteria can also be achieved with little, unstructured information.

6.4 Patterns across the main MaaS Ecosystems worldwide and their development since 2015 as a Result of Data Analysis and Interpretation

In order to examine the status quo and the development of the most important MaaS ecosystems worldwide, we analyzed 28 providers of 40 cities and thus also provider-city combinations with mentions of MaaS ecosystems according to our definition (in Fig. 6.3). We were able to identify the following patterns: first, patterns 1 and 2 in all mentions and then to patterns 3 to 7 in the mentions of the variables of the individual constructs of ecosystems (see also endnote 1).

Pattern 1: The total number of mentions of the selected providers and cities in the international German- and English-language press has increased more than tenfold with the development of MaaS by 2019 (Fig. 6.4a), until, in early 2020, it collapsed to less than half of mentions compared to the previous year. Since early 2020, mentions have increased again but have fallen by 46% compared to the peak in 2019. A possible explanation for this pattern could be the worldwide fear of corona virus contagion, the geopolitical challenges and the general economic climate since 2020. However, the development of mentions of MaaS ecosystems in newspaper articles and analyst reports could also follow a "hype cycle",[5], [14] according to which the visibility of a new technology initially rises sharply over time, then falls sharply and finally settles at a medium level.

Pattern 2: The number of mentions of value propositions, technical platform and also value drivers fell from 2015 to 2019 to 2023 (from 65% of mentions overall to 48%) in favor of mentions of the operating model and governance mechanisms (from 35% of mentions overall to 52%), see Fig. 6.4b. Therefore, the organization of actors, data and financial flows in the operating model through rules and regulations as part of governance appears to be gaining in importance.

Pattern 3: In terms of the value proposition, a shift toward individualized customer solutions can be observed (from 16 to 27% of mentions, Fig. 6.5a), while the integration of shared mobility solutions has been declining since 2019 (from 43% of mentions in 2019 to 28% in 2023), probably due to COVID-19. Additional benefits such as information about important events and ticketing in a region are becoming less important (decline from 34 to 20% of mentions).

Pattern 4: In terms of operating model, there has been a sharp decline in the importance of data flows over the years (from 31 to 8% of mentions, see Fig. 6.5b), but an increasing focus on financial flows (from 10 to 20% of mentions). In contrast, the importance of asset collaboration and asset investment, as well as partner (management) and relationships, has only increased slightly over the years (to 12%, 11%, 18% and 31% in 2023).

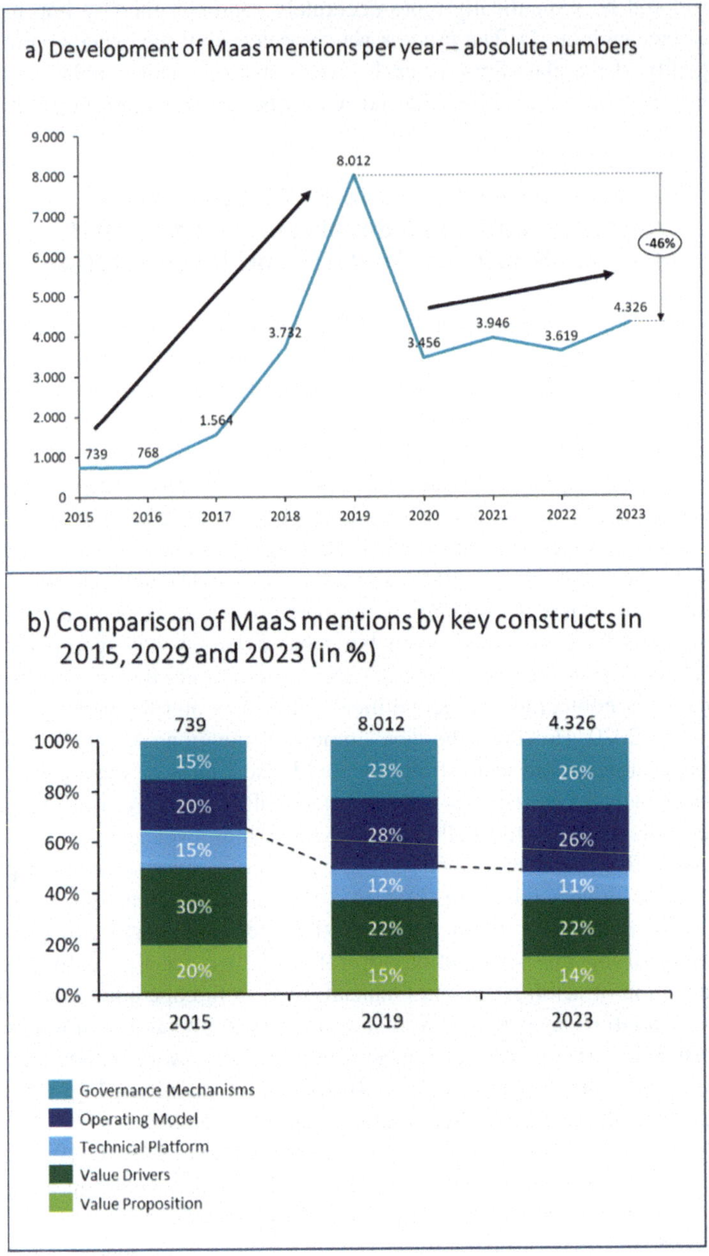

Fig. 6.4 Development of MaaS mentions a) per year and b) by key constructs 2015, 2019 and 2023. *Source* Own design, see also endnote 1

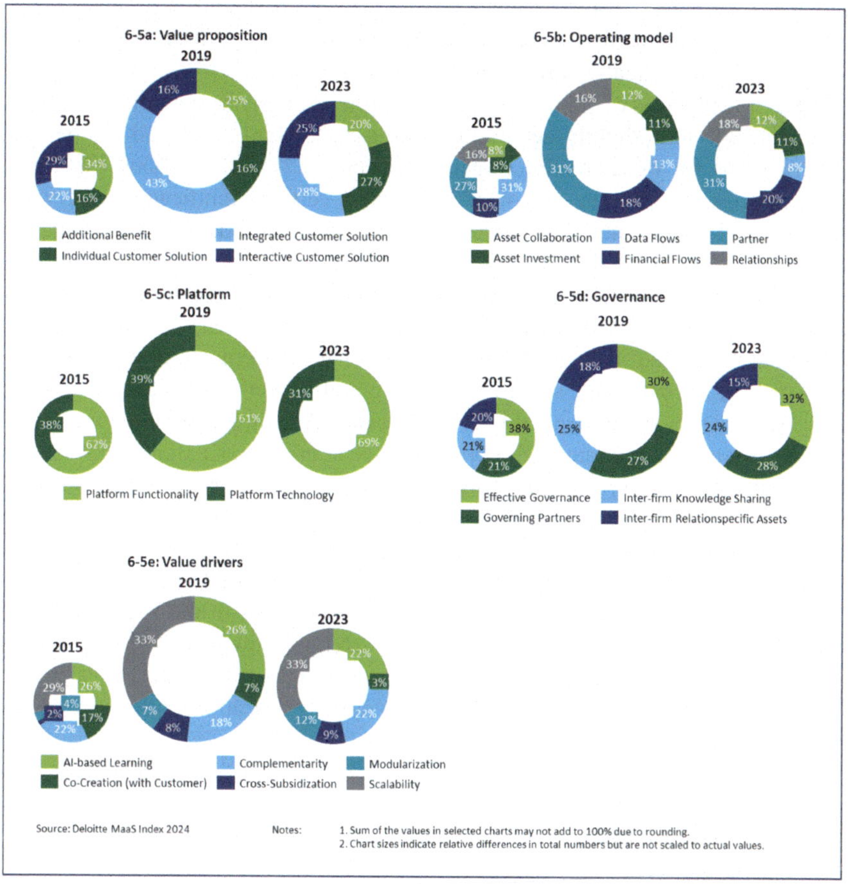

Fig. 6.5 Development of MaaS variables for each key construct. *Source* Own design, see also endnote 1

Pattern 5: Regarding the platform, there is further stability with a slight decrease in the importance of platform technology (from 38 to 31% of mentions) and a slight increase in the importance of platform functionality (from 62 to 69%, Fig. 6.5c). This means that a reliable technology stack for the seamless integration of services, data analytics and user experience is already largely in place and may have been driven by innovations in APIs, backend systems and data delivery.

Pattern 6: As MaaS ecosystems mature, the importance of governance increases (Fig. 6.4b), albeit relatively evenly distributed across the individual governance mechanisms and variables, as the number of mentions in the press remained roughly the same between 2015 and 2023. The importance of a basic governing of partners in the network increases only slightly (from 21 to 28% of mentions), while effective governance (rules, regulation, and role of partners) and relationship-specific values such as trust decrease slightly in importance (from 38 to 32% and from 20 to 15% of mentions respectively, see Fig. 6.5d).

Pattern 7: In terms of value drivers, the complementarity of partners, scalability in the network and (slightly decreasing) AI-based learning were and are particularly important for value creation (together constant at 77% of mentions, Fig. 6.5e). Value co-creation with the customer also had a certain significance as a value driver in 2015 due to the high importance of the value proposition at the time but is declining (from a combined 17% to just 3% of mentions). In contrast, the creation of interfaces to partners (modularization) and the cross-subsidization of weak partners are gaining in importance (from a combined 6% to 21% of mentions).

It should be noted that Fig. 6.5 only shows changes in mentions for the entire group of MaaS ecosystems, some of which are at different stages of development, partly because they were founded in different years. A comparison of the individual MaaS ecosystems as the analyzed provider-city combinations shows that it is in fact possible to identify such patterns across all companies because at one point in time (e.g., in the year 2023), the different ecosystems, whether founded earlier or just recently, hardly differ from one another and show fairly stable patterns. In addition, validation interviews with managers from 20% of the MaaS providers and cities surveyed in early 2024 confirm that our findings from the external data sources were in line with manager perceptions (average agreement of 5.6 on a scale from no agreement [25] to very strong agreement [38], see endnote 1),. There are also no major differences depending on the geographic location of the ecosystems on the world map.

6.5 Implications

Using our method of innovative text analysis, we were able to obtain patterns about the most important MaaS ecosystems worldwide and their development since 2015. Implications can be derived from this – on the one hand in Sect. 6.5.1 hints for improving MaaS ecosystems along the constructs and variables identified for the successful design of an ecosystem (Table 6.1 in Sect. 6.3.1) and on the other hand recommendations for more structural changes toward a successful design of transaction ecosystems for MaaS (Sect. 6.5.2)—not only, but also for traditional car manufacturers.

6.5.1 Hints for improving MaaS Ecosystems along its Constructs

As emphasized in the introduction, NLP as an "uncovering method" can be seen as a first step toward practical improvement (and also for in-depth theory building). The seven patterns in the development of MaaS ecosystems provide starting points for nine hints for practical improvements according to the criteria identified for the design of successful (MaaS) ecosystems (see endnote 1):

First hints for improving existing MaaS ecosystems could follows from patterns 1 and 7:

1. Create further media attention for MaaS in order to strengthen interactive customer solutions, which means an acceleration of the current trend.
2. From now on, take greater account of customer wishes and let customers dream in order to involve them more in the co-creation of MaaS values, which would be a trend reversal.

Two further hints for improving existing MaaS ecosystems could be derived from patterns 2 and 6:

3. Concentrate on organizing further the MaaS operating model and
4. Concentrate on implementing further the necessary governance mechanisms for MaaS (that both accelerate the current trend) to strengthen these ecosystems.

Two other hints for practical improvement of existing MaaS ecosystems can be derived from pattern 3:

5. Create a further individualized approach to MaaS customers to strengthen the individual customer solution, which also accelerates the current trend.
6. From now on, MaaS should be offered as a bundle of offerings with additional personalized benefits beyond transportation from A to B to strengthen the integration of further services and additional benefits such as comprehensive booking and billing functions or/and information about events, and weather conditions. This would be a trend reversal.

Moreover, two further hints for improving existing MaaS ecosystems are deduced from pattern 5:

7. Further focus on financial flows in MaaS ecosystems to ensure profitability, which accelerates the current trend.
8. From now on, data transfer within MaaS ecosystems must be improved. This would be a trend reversal.

As pattern 5 is stable, it only offers one further hint for improving existing MaaS ecosystems.

9. The creation of a technical platform can be somewhat removed from the focus of the broader ecosystem strategy once potential solutions are properly established.

Recommendations for more Structural Changes toward a successful design of Transaction Ecosystems for MaaS

These nine hints for practical improvements of existing MaaS ecosystems, derived from the patterns in MaaS ecosystem development, identified in Sect. 6.4, also provide recommendations for more structural changes toward a successful design of transaction ecosystems for MaaS (see also endnote 1)—not only, but also for traditional car manufacturers:

1. The successful design of further transaction ecosystems for MaaS cannot be achieved through Brokerage alone:

MaaS has so far often been offered as an app, i.e., as a platform through which the mobility providers involved in the ecosystem link their mobility services and through which the customer can obtain multimodal mobility from a single source. The provider of the app often has the role of an orchestrator [39] who coordinates the partners in the MaaS ecosystem. However, this orchestrator should not be a pure broker, i.e., not an app or platform-operating software company. Instead, it should be involved in the joint value creation in the ecosystem by providing mobility services. A pure app provider would have too little leverage to address customers in a more interactive, individualized and integrated way and to offer mobility bundles with additional benefits as seamlessly as possible from a single source, in accordance with hints 1, 5 and 6. Given their ability to orchestrate complex value chains, this would also allow traditional automotive companies, above all manufactures, to design and orchestrate MaaS offerings in transaction ecosystems in the new competition with car and ridesharing companies.

2. The successful design of further transaction ecosystems for MaaS must address the private and commercial sectors:

To be successful with a platform, according to hint 9, it is less the technology and functionality of the platform that is decisive than the "traffic" on the platform as a prerequisite for achieving scale and network effects [1, 12]. Thus, on the one hand, additional private customers should be addressed in a personalized manner, for example older people who are often willing to pay for individualized mobility. However, on the other hand, it also makes sense to open the platform to corporate customers, whose CO_2-reporting is increasingly being questioned and who can better appeal to new and, above all, younger talent with modern mobility offers. For example, they could be offered a mobility budget [26] that enables their employees to use MaaS. Such a mobility budget is a financial budget that employers make available to their employees as an additional benefit for their individual commuting and leisure trips. Addressing these additional, especially commercial customers is important because the other eight hints for practical improvements derived in the last subchapter are only worthwhile from a certain minimum size. The successful design of transaction ecosystems for MaaS must address not only private customers, but also commercial sectors. As traditional automotive companies, especially manufacturers, are able to address a mass market of both private and business customers, they could scale MaaS offerings in transaction ecosystems in the new competition with car and ridesharing companies.

3. The successful design of further transaction ecosystems for MaaS requires public–private Partnerships (PPP):

The implementation of the derived hints for practical improvements requires stronger PPP, [37] due to the central role of the public sector in the regulation of (urban) mobility and data management. Only with government help through new laws and newly created data interfaces can the main obstacle to the development of many MaaS ecosystems be removed: often insufficient data flow because such data is not shared between mobility providers or across city borders. This requires trust and transparency between the parties involved, which only the mobility providers themselves can create. However, regulations are also needed: National and supranational regulations that set environmental targets and timetables, but also regional support, e.g., through the creation of regional data interfaces that bundle and standardize mobility data in order to reduce the effort required for data integration within the company and increase the efficiency of the services based on it. Due to their political connections and influence, PPPs would also help traditional automotive companies, especially car manufacturers, to shape MaaS offerings in transaction ecosystems in the new competition with car and ridesharing companies.

With these hints for improving MaaS ecosystems and accelerating its development as well as with the recommendations for more structural changes toward a successful design of transaction ecosystems for MaaS, joint value creation in these MaaS transaction ecosystems (see Chap. 2) should work better. This is a prerequisite of additional engagement of private investment, e.g., from traditional automotive companies.

However, the question arises as to how MaaS and private car ownership fit together and where the traditional car manufacturers will remain when MaaS offers finally become more profitable and their spread is more driven by urban planners, so that they are offered more consistently and used more widely. This question will be addressed in the next chapter of this book.

Notes

1. From a technical perspective, it builds on recent advances in AI, such as the (supervised and unsupervised) training of deep neural networks. To avoid the risk of using pre-trained models and generic training data with unknown accuracy in the target domain, the construction of premises and hypotheses can be fine-tuned using a small number of positive and negative labels that can be manually extracted from the documents.

2. We have therefore defined that one of the following three conditions must be met for an article to be included and analyzed in the study: (I) the article refers to MaaS ecosystems or a known MaaS ecosystem provider or (II) the article discusses a linked multimodal offering or (III) the article mentions at least two collaborations.
3. Using this web application, an iterative, semi-supervised process was carried out (see endnote 1). First, the filtered documents from the previous step were tagged in the web tool for sentence-level labeling. The researchers thus assigned the interpreted variables to the corresponding sentences in the first labeling phase. Subsequently, the labeled sentences were used to form hypotheses in the NLI pipeline, with the aim of covering most of the previously labeled sentences. The NLI parameters were then manually fine-tuned to optimize quality. This semi-supervised approach provided us with customized NLI classifiers for each variable in each construct. Finally, in NLP-based classification, the classifiers are applied to the entire set of sentences in the filtered documents
4. During the iterative refinement process, the domain specialists manually validated more than 250 documents and identified a spectrum of positive and negative instances from all variables.

   These variables were then used to fine-tune the NLI parameters, and the underlying Artificial Neural Networks until the NLP pipeline provided adequate performance for the unique task for each variable. In a first round, as expected, due to the complex topic of MaaS, even in a combined AI and business researcher approach, six of the 22 considered variables failed to meet the required F1 score. Such variables would raise concerns about their significance and the quality of insights they can offer. Variables with a higher F1 score are included in the interpretation of results with greater significance.
5. The Hype Cycle is a graphical representation of the maturity, adoption and social application of specific technologies. It was developed by the technology research and consulting firm Gartner (Fenn, J., & Raskino, M. (2008). *Mastering the hype cycle: How to choose the right innovation at the right time.* Harvard Business Press).

   The Hype Cycle provides a visual and conceptual presentation of how a technology will evolve over time (Rogers, E. M. (2010). *Diffusion of innovations* (4th ed.). Simon and Schuster). This allows companies to better understand the promise of a technology and its potential future development, manage expectations, and mitigate risks (Linden, A., & Fenn, J. (2003). Understanding Gartner's hype cycles. *Strategic Analysis, 27*(1), 1–13). The Hype Cycle is divided into five phases: Technology Trigger, Peak of Inflated Expectations, Trough of Disillusionment, Slope of Enlightenment, and Plateau of Productivity.

References

1. Adner, R. (2017). Ecosystem as structure. *Journal of Management, 43*(1), 39–58.
2. Alpaydin, E. (2020). *Introduction to machine learning.* MIT Press.

3. Alyavina, E., Nikitas, A., & Tchouamou Njoya, E. (2020). Mobility as a service and sustainable travel behaviour: A thematic analysis study. *Transportation Research Part F: Traffic Psychology and Behaviour, 73,* 362–381.
4. Arkoudi, I., Krueger, R., Lima Azevedo, C., & Pereira, F. C. (2023). Combining discrete choice models and neural networks through embeddings: Formulation, interpretability and performance. *Transportation Research Part B: Methodological, 175,* Article 102783.
5. Bellinger, C., Sharma, S., & Japkowicz, N. (2012). One-class versus binary classification: Which and when? *2012 11th International Conference on Machine Learning and Applications, 2,* 102–106. IEEE.
6. van den Berg, V. A., Meurs, H., & Verhoef, E. T. (2022). Business models for Mobility as a Service (MaaS). *Transportation Research Part B: Methodological, 157,* 203–229.
7. Bernard, D., Burgstahler, D., & Kaya, D. (2018). Size management by European private firms to minimize proprietary costs of disclosure. *Journal of Accounting and Economics, 66*(1), 94–122.
8. Brandenburger, A. M., & Stuart, H. W. (2007). Biform games. *Management Science, 53*(4), 537–549.
9. Burchard, C. H., Proelss, J., Schäffer, U., & Schweizer, D. (2021). Bad news for announcers, good news for rivals: Are rivals fully seizing transition-period opportunities following announcers' top management turnovers? *Strategic Management Journal, 42*(3), 579–607.
10. Choi, J., Menon, A., & Tabakovic, H. (2021). Using machine learning to revisit the diversification-performance relationship. *Strategic Management Journal, 42*(9), 1632–1661.
11. Choudhury, P., Allen, R. T., & Endres, M. G. (2021). Machine learning for pattern discovery in management research. *Strategic Management Journal, 42*(1), 30–57.
12. Cusumano, M. A., Gawer, A., & Yoffie, D. B. (2019). *The business of platforms: Strategy in the age of digital competition, innovation, and power*. HarperCollins Publishers.
13. Cusumano, M. A., Gawer, A., Yoffie, D. B., von Bargen, S., & Acquay, K. (2023). *Is there a "platform premium"? An exploratory study of unicorn business models and valuations.* Retrieved January 29, 2007, from https://questromworld.bu.edu/platformstrategy/wp-content/uploads/sites/49/2023/06/PlatStrat2023_paper_5.pdf
14. Dedehayir, O., & Steinert, M. (2016). The hype cycle model: A review and future directions. *Technological Forecasting and Social Change, 108,* 28–41.
15. Dušek, R., Galias, C., Wojciechowska, L., & Wawer, A. (2023). Improving domain-specific retrieval by NLI fine-tuning. *2023 18th Conference on Computer Science and Intelligence Systems (FedCSIS),* 949–953.
16. Dyer, J. H., Singh, H., & Hesterly, W. S. (2018). The relational view revisited: A dynamic perspective on value creation and value capture. *Strategic Management Journal, 39*(12), 3140–3162.
17. Eisenmann, T., Parker, G., & Van Alstyne, M. (2006). Strategies for two-sided markets. *Harvard Business Review, 84*(10), 92–101.
18. Engelberg, J., McLean, R. D., & Pontiff, J. (2018). Anomalies and news. *The Journal of Finance, 73*(5), 1971–2001.
19. Facin, A. L. F., Vasconcelos Gomes, L. A. de, Mesquita Spinola, M. de, & Salerno, M. S. (2016). The evolution of the platform concept: A systematic review. *IEEE Transactions on Engineering Management, 63*(4), 475–488.
20. Gerwanski, J. (2020). Managers' incentives and disincentives to engage with integrated reporting or why managers might not adopt integrated reporting: An exploratory study in a nascent setting. *Qualitative Research in Accounting & Management, 17*(4), 553–587.
21. Gharehchopogh, F. S., & Khalifelu, Z. A. (2011). *Analysis and evaluation of unstructured data: Text mining versus natural language processing.* 2011 5th International Conference on Application of Information and Communication Technologies (AICT), 1–4.
22. Gubelmann, R., Kalouli, A. L., Niklaus, C., & Handschuh, S. (2023). When truth matters—addressing pragmatic categories in natural language inference (NLI) by large language models (LLMs). *Proceedings of the 12th Joint Conference on Lexical and Computational Semantics (SEM 2023),* 24–39.

23. Guidon, S., Wicki, M., Bernauer, T., & Axhausen, K. (2020). Transportation service bundling—For whose benefit? Consumer valuation of pure bundling in the passenger transportation market. *Transportation Research Part A: Policy and Practice, 131,* 91–106.
24. Hagiu, A., & Rothman, S. (2016). Network effects aren't enough. *Harvard Business Review, 94*(4), 64–71.
25. Hietanen, S. (2014). Mobility as a Service—The new transport model? *Eurotransport, 12*(2), 2–4.
26. Ho, C. Q. (2022). Can MaaS change users' travel behaviour to deliver commercial and societal outcomes? *Transportation Research Part A: Policy and Practice, 165,* 76–97. https://doi.org/10.1016/j.tra.2022.09.004
27. Iansiti, M., & Lakhani, K. R. (2020). *Competing in an age of AI: Strategy and leadership when algorithms and networks run the world.* Harvard Business Review Press.
28. Jacobides, M. G., Cennamo, C., & Gawer, A. (2018). Towards a theory of ecosystems. *Strategic Management Journal, 39*(8), 2255–2276.
29. Jeppe, A., Anzani, A., Handte, M., Proff, H., & Marrón, P. J. (2025). *Patterns in the development of Mobility as a Service ecosystems: An AI based approach in a limited and unstructured data environment.* Under review.
30. Kalouli, A. L., Crouch, R., & De Paiva, V. (2020). Hy-NLI: A hybrid system for natural language inference. *Proceedings of the 28th International Conference on Computational Linguistics,* 5235–5249.
31. Kamargianni, M., & Matyas, M. (2017). *The business ecosystem of Mobility-as-a-Service.* 96th Transportation Research Board (TRB) Annual Meeting, Washington, DC, United States.
32. Karlsson, I., Mukhtar-Landgren, D., Smith, G., Koglin, T., Kronsell, A., Lund, E., Sarasini, S., & Sochor, J. (2020). Development and implementation of Mobility-as-a-Service – A qualitative study of barriers and enabling factors. *Transportation Research Part A: Policy and Practice, 131,* 283–295.
33. Katz, M. L., & Shapiro, C. (1985). Network externalities, competition, and compatibility. *The American Economic Review, 75*(3), 424–440.
34. Lewrick, M. (2021). *Business Ökosystem Design.* Vahlen.
35. Li, J., Sun, A., Han, J., & Li, C. (2020). A survey on deep learning for named entity recognition. *IEEE Transactions on Knowledge and Data Engineering, 34*(1), 50–70.
36. Mayring, P. (2014). Qualitative content analysis: Theoretical background and procedures. In A. Bikner-Ahsbahs, C. Knipping, & N. Presmeg (Eds.), *Approaches to qualitative research in mathematics education* (pp. 365–380). Springer.
37. Polydoropoulou, A., Pagoni, I., Tsirimpa, A., Kamargianni, M., & Tsouros, I. (2020). Prototype business models for Mobility-as-a-Service. *Transportation Research Part A: Policy and Practice, 131,* 149–162.
38. Rochet, J.-C., & Tirole, J. (2003). Platform competition in two-sided markets. *Journal of the European Economic Association, 1*(4), 990–1029.
39. Sjödin, D., Parida, V., Palmié, M., & Wincent, J. (2021). How AI capabilities enable business model innovation: Scaling AI through co-evolutionary processes and feedback loops. *Journal of Business Research, 134,* 574–587.
40. Smith, T. (2024). *Mobility startup MaaS Global files for bankruptcy.* Tech.eu.
41. Tabassum, A., & Patil, R. R. (2020). A survey on text pre-processing & feature extraction techniques in natural language processing. *International Research Journal of Engineering and Technology, 7*(6), 4864–4867.
42. Tidhar, R., & Eisenhardt, K. M. (2020). Get rich or die trying…finding revenue model fit using machine learning and multiple cases. *Strategic Management Journal, 41*(7), 1245–1273.
43. United Nations. (2015). *Paris Agreement.* UNFCCC.
44. Vargo, S. L., & Lusch, R. F. (2004). Evolving to a new dominant logic for marketing. *Journal of Marketing, 68*(1), 1–17.
45. Zhao, X., Vaddadi, B., Sjöman, M., Hesselgren, M., & Pernestål, A. (2020). Key barriers in MaaS development and implementation: Lessons learned from testing Corporate MaaS (CMaaS). *Transportation Research Interdisciplinary Perspectives, 8,* Article 100227.

Future Coexistence of Mobility as a Service and Private Car Ownership as well as the Role of Traditional OEMs

7

Harald V. Proff, Nico Zauner, and Heike Proff

Contents

Abstract

This Chapter examines the future coexistence of Mobility as a Service and private car ownership. This raises the question of the future role of traditional OEMs. Will they continue to coexist with MaaS offerings, will they even be displaced by them, or will they join forces with them in larger ecosystems in a coopetition relationship?

H. V. Proff (✉) · N. Zauner
Deloitte, Düsseldorf, Germany
e-mail: hproff@deloitte.de

N. Zauner
IESE Business School, Barcelona, Spain

H. Proff
University of Duisburg-Essen, Duisburg/Essen, Germany
e-mail: heike.proff@uni-due.de

H. Proff et al. (eds.), *Future of the Automotive Industry*, Future of Business and Finance, https://doi.org/10.1007/978-3-032-26308-7_7

7.1 Introduction

Chapter 6 analysed the status quo of MaaS offerings from transaction ecosystems worldwide and identified possible starting points for acceleration. But where does this leave traditional car manufacturers? Will they continue to coexist with MaaS offerings, will they even be displaced by them, or will they join forces with them in larger ecosystems in a coopetition relationship (see Chap. 2 in this book)?

When talking about the future of the automotive industry in the context of MaaS solutions, it is usually implicitly assumed that people will either own their own car or use different modes of transport such as bus, train, car sharing or e-scooters flexibly and in combination – i.e. multimodally – as required MaaS, i.e. opt for one of the two mobility solutions [16]. For this reason, car companies pay little attention to these multimodal mobility services because MaaS is only an option for a smaller group of their customers anyway – namely those who live in urban centres. On the other hand, when calculating potential customers' willingness to pay, MaaS providers often assume that they do not own a car [14].

However, viewing the private car and MaaS as an either-or proposition is one of the reasons why MaaS solutions have not been very successful to date (see Chap. 6 in this book) and why car OEMs are missing out on potential sales. New opportunities, but also new challenges are likely to arise in the future as the product life of vehicles is expected to increase due to environmental reasons. Besides, manufacturers will actively look after their vehicles for more than the approximate four years that they have generally had to date, before they are transferred to the so-called independent (non-branded) aftermarket – i.e. to the care of independent workshops or third-party service providers [5]. For car manufacturers, this means that not only will car sales grow less as a result of more mobility services, but they will also have to service each car for longer than before. They must therefore generate more revenue per vehicle—with new sales models and new customer interaction models such as leasing, subscription or long-term usage contracts that complement the new mobility services on the market. Instead of "either your own car or mobility services", a "both-and" is therefore the better solution for them too.

The combination of private car and MaaS fits well with mobility concepts focusing on multimodality – i.e. the flexible combination of modes of transport such as public transport, bicycles, sharing services, rail and even the private car. In view of climate issues, it would be beneficial if the private car was an electric vehicle that runs on green energy. However, CO_2 emissions can already be significantly reduced if a private combustion-powered car only travels half of the current average yearly mileage by increasing the use of multimodal mobility options.

In addition to the powertrain transition, which includes the switch from combustion engines to more environmentally friendly drive systems such as electric mobility, the transport transition is also becoming increasingly important for car manufacturers. It aims to make the entire transport system with private cars and MaaS more sustainable and flexible, which is particularly relevant for car OEMs when it comes to offering customized leasing, subscription or long-term usage contracts. This chapter therefore examines the coexistence of MaaS and private vehicle ownership in more detail. First, the current situation of this relationship is analysed (Sect. 7.2), before indications for the future coexistence of MaaS and private vehicle ownership are presented: The results of the Deloitte Global Automotive Consumer Study 2025 on expected changes in customer behaviour up to 2035 are presented, which justify a demand impulse ("market pull") for further coexistence, but with stronger links (Sect. 7.3). The results of other Deloitte studies then show a slow change in market segmentation, which justifies a market push towards further coexistence of vehicle ownership and MaaS and, in doing so, towards greater cooperation between traditional car manufacturers and MaaS providers (Sect. 7.4). Finally, the chapter ends with a brief outlook (Sect. 7.5).

7.2 Current Relationship between MaaS and Private Car Ownership

There are not many studies that analyse the relationship between MaaS and private car ownership. As already emphasized, it is generally assumed that the relationship is "either-or".

For example, studies on attitudes towards MaaS [16, 2] assume a correlation according to which owning a car decreases with an improved MaaS system. In addition, caution is required in surveys, particularly in the area of (individual) mobility, due to the large deviations between the attitudes surveyed and actual behaviour. The problems with existing MaaS solutions highlighted in Chap. 6 of this book make it clear that they are often based on unrealistic assumptions and reflect political wishful thinking rather than actual reality.

Ultimately, the existing studies show the same picture as Storme et al. (2020) [23]: primarily urban, mobility-savvy test subjects are recruited for the study, mostly students or members of university institutions. Around a third of the participants discontinue the trial prematurely as participation is not compatible with their individual everyday requirements. Of the remaining two thirds, around 80% reduce their individual car use to varying degrees. The remaining 20%, on the other hand, maintain their previous mobility behaviour with intensive use of their private car [23]. Among the urban academics with a particular affinity for MaaS, around half of the participants are thus reducing their use of private vehicles as a result of improved MaaS services [23, 11]. It can be assumed that corresponding studies in suburban or rural areas would lead to significantly different results, as individual mobility, especially by private car, is much more important there.

However, urban vehicle owners only account for around 15% of a car manufacturer's total sales [6]. Against this background, traditional car manufacturers justify their lack of interest in developing mobility solutions.

However, this argumentation appears increasingly unjustified. In view of the sharp rise in vehicle prices since the start of the Covid-19 pandemic and the subsequent inflation phase (in some cases by up to 40% within four years), customers are beginning to adjust their purchasing behaviour. One indicator of this is the decoupling of the development of global GDP and global vehicle sales, which were closely linked until the start of the pandemic, and whose previous parallelism no longer exists: vehicle prices are now rising much faster [12].

Against this backdrop, the top management of car manufacturers and major suppliers are finding it increasingly difficult to present a convincing growth story to shareholders and the capital markets – particularly regarding sales. This is particularly unfavourable in an industry such as the automotive industry, which is not one of the capital markets' preferred investment targets due to its capital-intensive structures [20]. Providers are increasingly endeavouring to offer subscription models, longer vehicle holding periods and mobility budgets instead of traditional company cars (see Sect. 7.4) in order to send clear impulses to the capital markets. After the hype and subsequent de-hype surrounding "smart mobility" solutions in the mid to late 2010s, they are now also enabling car manufacturers to take a more pragmatic, realistic approach to new mobility solutions.

The opening of car manufacturers and suppliers to MaaS solutions is also driven by the availability of and control over proprietary data. This is because both are a key issue for all car OEMs when competing with technology companies. This is why data availability and protection form the basis for all smart mobility concepts. In view of the increasing challenges posed by tech companies and the growing influence of artificial intelligence (AI), this aspect is now taking on a new urgency. Here, too, the "ecosystem play" discussed in this book is gaining in importance, in which governance mechanisms must define common rules and regulations for data collection and data transfer between the partners in a (here transactional) ecosystem with coopetition and create mutual trust (see Chap. 2).

It is therefore necessary to redefine the coexistence of MaaS and private vehicle ownership to generate new market opportunities and advantages through cooperation. This need is likely to be driven by market pull from expected changes in demand behaviour as well as market push from changes in market segmentation. This will be shown in the next two sections.

7.3 Market Pull for a Coexistence of MaaS and Private Car Ownership – Expected Changes in Customer Behaviour up to 2035

For the latest Deloitte Global Automotive Consumer Study 2025 [3], 33,000 consumers from 30 countries across the world were surveyed in October 2024 regarding their views on the relationship between private vehicle usage and MaaS. The survey was conducted online, with identical questionnaires translated into the respective national language [3]. The respondents were also asked about the proportion of their mobility needs that they currently (want to) cover with their private car and with MaaS systems in 2030. Three customer groups were distinguished for better differentiation:

- "car enthusiasts", who cover between 100% and 80% of their mobility needs with their own cars and only use MaaS between 0% and 20%,
- "car avoiders", who only cover between 0% and 20% of their mobility needs by car and therefore use MaaS systems for between 80% and 100%, and
- in between, "mobility pragmatists", who cover between 20% and 80% of their mobility needs either with an individual car or with MaaS systems [3].

While the needs of the "car enthusiasts" are already largely covered and there are only limited business opportunities with the "car sceptics", the strategic focus of automotive companies should be on increasing sales with the "mobility pragmatists". This target group offers the greatest potential for sustainable growth and long-term market gains. Therefore, MaaS providers should better understand the "mobility pragmatists" in order to generate more "traffic" on their platform to become profitable and reach the next stage of the emergence of a MaaS ecosystem (see Chap. 6). Traditional car manufacturers should also focus on the "mobility pragmatists" to tap into significant market potential.

It should therefore be the strategic goal of traditional car manufacturers and MaaS providers to offer "mobility pragmatists" private car use in conjunction with MaaS solutions instead of continuing to develop only "nice-to-have" solutions for "car sceptics" who are already convinced by MaaS solutions. The focus should be on creating added value for those who take a pragmatic approach to their mobility design and create a market pull with their needs in an increasingly diversified mobility market.

The results of the Deloitte's Global Automotive Consumer Study 2025 in Table 7.1 show that the "car enthusiasts and the car avoiders" are more strongly represented in Germany than in the overall sample, but that the group of "car pragmatists" is increasing over time in both samples. Around 58% of respondents in Germany will be classified as "car enthusiasts" and 15% as "car avoiders" in 2025, meaning that there will only be 27% "mobility pragmatists". By 2030, their share will rise to 32%. In the global sample, the proportion of mobility pragmatists rises from 35% to 41%.

Looking at demographic characteristics such as age, income and place of residence (urban or rural), it becomes clear that the age effect accounts for the most pronounced difference of all the parameters analysed. In particular, there is a striking difference between the under-34 age group and the over-55 s. In Germany, only 14% of respondents over the age of 55 will be "mobility pragmatists" in 2025, compared to 53% of those under 34. In the global sample, 24% of respondents over the age of 55% and 47% of those under 34 are in the "mobility pragmatists" group.

Table 7.1 Use of private cars and MaaS in Germany and worldwide. *Source* Deloitte, 2025

		Germany			Global		
		Car-enthusiasts	Mobility pragmatists	Car-avoider	Car-enthusiasts	Nobility-pragmatists	Car-avoider
Ø Total 2025		58%	27%	15%	52%	35%	13%
Ø Total 2030		52%	32%	16%	45%	41%	13%
Ø 2025	Younger (<34 y.o.)	39%	53%	8%	44%	47%	9%
	Older (>55 y.o)	67%	14%	18%	59%	24%	17%
	High income	62%	31%	7%	53%	37%	10%
	Low income	45%	27%	28%	51%	30%	18%
	Urban	53%	30%	18%	48%	39%	13%
	Suburban/rural	62%	25%	12%	58%	29%	13%
Ø Male		61%	25%	13%	54%	34%	11%
Ø Female		54%	29%	17%	50%	35%	15%

The difference between urban citizens and respondents living in suburban/rural areas was surprisingly small. In Germany, only 30% of urban citizens and 25% of suburban/rural respondents were "mobility pragmatists" (in the overall sample 39% and 29% respectively). This result underpins the findings of the study by Storme et al. 2020 [23] that even in urban life the private car is still used intensively. Income and gender also show only minor differences between German and global mobility pragmatists (see Table 7.1).

A comparison of the results for 2025 for America, Europe and Asia shows the lowest number of "mobility pragmatists" in America (29%), followed by the European countries with 32%, while their share in the Asian countries is already at 42%. This order remains constant until 2030, but all regions show an increase in "mobility pragmatists" to 34% in America, 38% in Europe and 48% in Asia. With the exception of Asia, the greatest difference between young and older respondents can also be observed among the "mobility pragmatists"; the difference between residents of urban and suburban/rural areas is slightly higher among car enthusiasts (see Table 7.2).

Table 7.2 Use of private cars and MaaS—in America, Europe and Asia. *Source* Deloitte, 2025

		Americas			Europe			Asia		
		Car enthusiasts	Mobility pragmatists	Car avoider	Car enthusiasts	Mobility pragmatists	Car avoider	Car enthusiasts	Mobility pragmatists	Car avoider
Ø Total 2025		58%	29%	14%	54%	32%	14%	47%	42%	11%
Ø Total 2030		54%	34%	12%	47%	38%	14%	40%	48%	12%
Ø 2025	Younger (<34 y.o.)	46%	42%	12%	45%	46%	9%	40%	51%	8%
	Older (>55 y.o.)	68%	16%	16%	62%	21%	17%	40%	43%	17%
	Urban	54%	32%	14%	50%	35%	15%	38%	51%	11%
	Suburban/	64%	24%	13%	60%	28%	12%	44%	43%	14%

The survey clearly shows that for both automotive companies and MaaS providers, the key to additional growth comes from market pull from younger customers ("young drivers"). In addition, older customers ("silver agers") in ageing societies, such as in Germany and Japan, represent a growing customer group with purchasing power [10, 9], which will also trigger a market pull. Car manufacturers, particularly from Germany, are currently primarily tapping into the market potential of their traditional new car customers with an average age of 54. They will have to increasingly turn to younger target groups ("young drivers") and older customer segments ("silver agers") in order to tap into new market potential (see Fig. 7.1).

Figure 7.1 illustrates the untapped market potential of traditional, particularly German, car manufacturers in the form of a bell-shaped normal distribution across the age group of 18- to 80-year-old customers, with demand peaking at the age of 54. The figure thus clearly shows previously untapped market potential at the edges of this bell-shaped curve among younger and older people, the development of which is vital for car manufacturers in the current transformation of mobility towards more connected, shared, but also highly automated or autonomous mobility. This is because their needs will increasingly trigger a demand impulse ("market pull"] that MaaS providers and tech players would otherwise fulfil on their own.

Only a small proportion of the "young generation", i.e. people aged between 18 and 25, are still "young drivers". Young people today are no longer as focused on car use as older mobility customers with often habitualized mobility behaviour [15], partly because the importance of the car as a status symbol is declining among the younger generation and acceptance of new technologies is high [20, 7, 13]. They are more receptive to innovative MaaS offers [22]. Therefore, there is a high market potential among young people that traditional car manufacturers are not tapping into enough. Students in urban areas in particular, who will play a major role as future consumers with purchasing power [14, 22], [24] are turning away from cars and need alternative offers. This is likely to be a trigger for the aforementioned demand impulse ("market pull"). If traditional car manufacturers do not succeed in retaining more young customers by integrating their vehicles with the existing MaaS solutions into comprehensive mobility services, they could be pushed back into niche markets by car-, bike- and ridesharing or MaaS providers in the future (see also Chap. 1 in this book).

Populations are getting older, and older generations are becoming increasingly mobile [19]. For them, mobility means independence, social participation, physical activity and an increased quality of life [18]. However, older drivers are also much more likely than younger drivers to have problems in road traffic, especially in complex traffic situations such as turning left in oncoming traffic or merging and changing lanes on motorways at high speed. They are more affected by visual changes such as fog, rain or twilight and specific cognitive functions experience an age-associated reduction in performance[21]. ADAS, such as speed and distance control systems, lane departure warning systems, emergency braking and warning systems and parking aids, not only increase driving comfort, but can also compensate for the limitations of older drivers and ensure safe mobility into old age. They are seen as steps in the digital transformation towards autonomous driving (i.e. up to level 5 "no driver"), which would be the perfect mobility solution for older people [8]. Although there is a need to support older drivers, driver assistance systems have not yet been established on the market with the expected success. The market potential for traditional car manufacturers has hardly been exhausted (see Fig. 7.1).

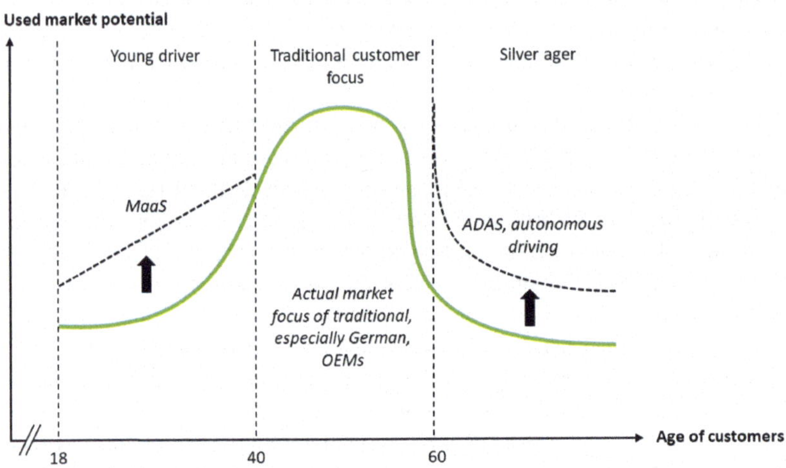

Fig. 7.1 Untapped market potential for traditional (especially German) car manufacturers. *Source* Own design based on Günther et al. 2021 and Günthner, Proff 2021

Studies at the University of Duisburg-Essen (see endnote 2) [10, 9] show that older drivers only use driver assistance systems that they trust and that offer a high level of user-friendliness. The willingness to pay for such assistance systems is higher in the group of 60 to 69-year-old drivers than among younger (50 to 59-year-olds] and even older drivers (over 70-year-olds). They are also likely to help trigger the demand impulse ("market pull") in future. If traditional car manufacturers fail in retaining these older customers by offering driver assistance systems on the road to autonomous driving combined with MaaS solutions, tech providers—with increasing vehicle automation in combination with the MaaS providers – could siphon off this market potential and push traditional car manufacturers back into niche markets in the future (see also Chap. 1 in this book).

7.4 Market Push of a Coexistence of MaaS and Private Car Ownership – Changes in Market Segmentation

In addition to demand stimuli from customers, further Deloitte studies [4, 3] also suggest a market push by providers towards a stronger coexistence and even combination of MaaS and private vehicle ownership.

The evidence shows that automotive companies are slowly breaking down the segmentation of their markets from the A-segment (subcompact cars) to the F-segment (luxury vehicles) and driving segments that complement MaaS solutions. This trend began with the CASE revolution.

However, automotive companies are also responding to the changes in customer ownership and usage structures (see last Sect. 7.5): More and more people are choosing individual mobility without owning a car, especially in Europe, North America and Japan – but with a time lag also in China, where, however, owning a car is still much more important as a place of retreat. Younger and older age groups in particular are increasingly using different forms of mobility. However, private customers in the current market focus of traditional (especially German) automotive companies (Fig. 7.1) are also increasingly leasing and renting vehicles. At the same time, companies can offer their employees car subscriptions instead of company cars. In Europe, they can also provide them with public transport tickets or mobility budgets (financial budgets that people can use flexibly for their individual mobility [15]). This is possible because the providers create these offers as part of the dissolution of their market segments.

The dissolution of traditional market segments and the creation of new offerings is being accelerated by the transition to electromobility, because a parallel offering of vehicles with combustion engines, plug-in hybrid vehicles and purely battery electric vehicles (BEVs) is only possible with a reduced product portfolio in which traditional segments are eliminated. As a result, there are more and more crossover vehicles, e.g. smaller but expensive sport utility vehicles (SUVs), across the traditional segments and body types.

Instead of the traditional segmentation of the automotive markets with vehicle segments A to F, traditional car manufacturers expect the entire mobility market to be segmented into five segments in future (Fig. 7.2): (1) privately owned vehicles (paid for in cash or financed), (2) vehicle leasing (mainly in fleets), (3) vehicle subscriptions (subscription contracts or subscriptions), (4) vehicle rental and (5) car and ridesharing [3, 4].

Studies by Deloitte surveying 1,500 global car buyers in 2022 [4] and 33,000 customers of cars and light commercial vehicles [3] show that customers can be assigned to these new market segments (Fig. 7.2). The results of the studies were based on global vehicle sales in the traditional market segments in 2017 (85 million units) and the vehicle sales forecast for 2030 and 2035 using our forecast models (88 and 85 million units worldwide). It is clear that the importance of privately owned vehicles – paid for in cash or financed – is declining. Leasing and rental vehicles, for example, will maintain their share and sharing and subscription models will increase in importance.

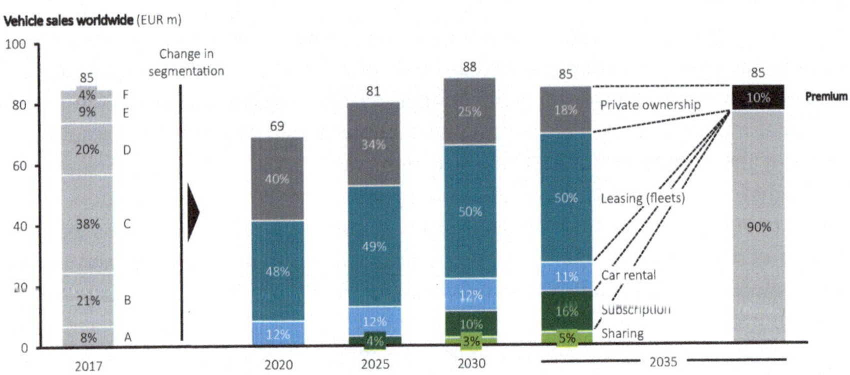

Fig. 7.2 Expected development of global vehicle sales up to 2025 and expected changes in the changes in the distribution of the new vehicle segments. *Source* Own design based on (Deloitte 2023 and 2024, [3, 4])

For traditional car companies, especially German premium manufacturers, the increasing importance of the market segments of vehicle leasing, vehicle subscriptions, vehicle rental and car and ridesharing means that the vehicles themselves are at risk of becoming more and more standardized, interchangeable goods ("commodities"). However, this risk seems low for the time being because no widespread breakthrough of undifferentiated, self-driving public robotaxis is expected before 2035 (even if they would not only be helpful for older people, see Sect. 7.3). Until then – and presumably even then – cars will still be long-lasting consumer goods that are selected by customers, even if market segmentation changes. The choice of vehicle therefore remains a choice of models, brands and prices. Even in the leasing, subscription, rental and car and ridesharing segments, there will still be a premium segment that opens up market opportunities for traditional (German) premium providers. And even with changing offers of employee mobility by corporate customers, income differences will not be levelled out—neither for the remaining privately owned vehicles nor for leased vehicles in vehicle fleets, which will account for around half of global vehicle sales by 2035. It is expected that the premium market will have a share of around 10% of the total market across all sub-segments by 2035. That would be 8.5 million vehicles).

However, premium needs to be redefined, especially in contrast to the offerings of Chinese premium providers. They are also claiming a new definition of premium through connectivity, infotainment as well as smart and autonomous mobility. Traditional (German) premium providers must therefore be faster and more creative in their definition of premium than their Chinese software-oriented competitors.

However, the current market push towards offering vehicles to private individuals and at the same time as part of mobility solutions by changing market segmentation is only just beginning. One of the reasons for this is that it fundamentally complicates the sale of vehicles and that sharing and subscription models are not yet really economically competitive (see Chap. 6 in this book). For this reason, traditional car manufacturers need to further develop the coexistence of the offer to private vehicle owners and for MaaS solutions and link the two more closely. Rental cars could become part of smart mobility solutions through long-term hire. They must then be offered on distribution networks in which various mobility providers are involved. Leased vehicles could be transferred to hybrid use to monetize the ever-increasing investment costs, in which vehicles are assigned to a private customer but are also integrated as sharing vehicles as part of a smart mobility system. What is already possible in private car parks [1][1] or real estate (AirBnB) [8] will also become established in the car industry in the future. When charging electric vehicles, especially when using vehicle batteries as energy storage systems or even fleet power plants [17], other partners such as energy suppliers, grid operators and charge point operators will be included in the smart mobility systems.

A stronger link between private vehicle ownership and MaaS offerings in turn requires the merger of car manufacturers in ecosystems as a fixed alignment of more than two partners on an overarching value proposition [1], accompanied by governance mechanisms and value drivers (see Chaps. 2 and 6 in this book).

Such transaction ecosystems thus become an enabler for a redefined relationship of coopetition between private cars and MaaS (see Chap. 2 in this book).

7.5 Outlook

In this chapter, it was explained that, on the one hand, holistic mobility services are in demand primarily from younger and older customers and, on the other hand, there is a new market segmentation focusing on vehicle leasing, subscription, vehicle rental and car- and ridesharing by the mobility providers. This means that not only will traditional car manufacturers and MaaS providers coexist, but there will have to be increased cooperation with competitors (coopetition) in mobility ecosystems.

These ecosystems enable social and ecologically sustainable mobility services that can improve the quality of life for all people in cities and rural areas. With increasing demand (more "traffic" on mobility platforms) in cooperation between predominantly private car and sharing providers and public local public transport (in private–public partnerships), they will also become profitable and thus, economically sustainable.

This means that the future of the automotive industry will not only be determined technologically by autonomous driving and the software (of a future SDV) (see Chaps. 4 and 5 in this book) but will also be decided on the sales side by bundling the offerings into holistic customer solutions. And that is good news for German automotive companies, which still need to build up technological expertise in autonomous driving functions and software in the face of new players like tech companies and Chinese competitors. This is because the new players have so far struggled on the sales side, particularly in Europe and Germany. The traditional car manufacturers have a head start here that needs to be maintained. However, this can only be achieved through stronger cooperation in transaction ecosystems.

Note

1. Parking platforms that offer normal parking on hourly basis within a private parking spot.

References

1. Adner, R. (2017). Ecosystem as structure: An actionable construct for strategy. *Journal of Management, 43*(1), 39–58.
2. Alyavina, E., Nikitas, A., & Tchouamou Njoya, E. (2024). Mobility-as-a-Service and unsustainable travel behaviour: Exploring the car ownership and public transport trip replacement side-effects of the MaaS paradigm. *Transport Policy, 150,* 53–70.

3. Deloitte. (2025). *Global automotive consumer study.*
4. Deloitte. (2023). *Future of Automotive Mobility Study.*
5. Deloitte. (2023). *Pathway to net-zero: Mastering the twofold goal of decarbonisation and profitability.* Deloitte Automotive. https://image.marketing.deloitte.de/lib/fe3111707564 0474771d75/m/2/f54ec435-704a-4092-8921-2011be41c998.pdf
6. Deloitte. (2021). *Nachfrage sucht Angebot.*
7. Elder, G. H. (1994). Time, human agency, and social change: Perspectives on the life course. *Social Psychology Quarterly, 57*(1), 4–15.
8. Guttentag, D., & Smith, S. (2017). Assessing Airbnb as a disruptive innovation relative to hotels: Substitution and comparative performance expectations. *International Journal of Hospitality Management, 64,* 1–10.
9. Günthner, T., & Proff, H. (2021). On the way to autonomous driving: How age influences the acceptance of driver assistance system. *Transportation Research Part F: Traffic Psychology and Behaviour, 81,* 586–607.
10. Günthner, T., Proff, H., Jovic, J., & Zeymer, L. (2021). Tapping into market opportunities in aging societies: The example of advanced driver assistance systems in the transition to autonomous driving. *International Journal of Automotive Technology and Management, 21*(1/2), 75–98.
11. Henshler, D. A., Ho, C. Q., & Reck, D. J. (2021). Mobility as a service and private car use: Evidence from the Sydney MaaS trial. *In Transportation Research Part, 1*(145), 17–33.
12. Heyn, M. (2025, March 25). *Presentation at the Mobility Innovation Summit of the VDA in the Strategy Session.* Berlin.
13. Hunecke, M., Groth, S., & Wittowsky, D. (2020). Young social milieus and multimodality: Interrelations of travel behaviours and psychographic characteristics. *Mobilities, 15*(3), 397–415.
14. Kraus, L., Proff, H., & Giesing, C. (2023). Composition of a mobility as a service offer for university students based on willingness to pay and its determinants. *International Journal of Automotive Technology and Management, 23*(2/3), 227–256.
15. Leontaris, T., Kraus, L., Willen, S., & Kleemann, F. (2024). Rediscovering everyday mobility - multimodal and sustainable travel thanks to mobility budgets? In H. Proff (Ed.), *Next chapter in mobility* (pp. 649–665). Springer Gabler.
16. Liljamo, T., Liimatainen, H., Pöllänen, M., & Utriainen, R. (2020). People's current mobility costs and willingness to pay for mobility as a service offerings. *Transportation Research Part A: Policy and Practice, 136,* 99–119.
17. Meyer, D., Kraus, L., Husemann, L., & Proff, H. (2025). Willingness of employees to accept bidirectional charging at the workplace. *Transportation Research Part D: Transport and Environment, 139,* Article 104517.
18. Musselwhite, C. (2018). Mobility in later life and wellbeing. In M. Friman, C. Ettema, & L. Olsson (Eds.), *Quality of life and daily travel* (pp. 235–251). Springer.
19. Nobis, C., Kuhnimhof, T., Follmer, R., & Bäumer, M. (2019). *Mobility in Germany—Time series report 2002–2008–2017.* Study by infas, DLR, IVT and infas 360 on behalf of the Federal Ministry of Transport and Digital Infrastructure. FE-No. 70.904/15.
20. Proff, H. (2019). *Multinational automotive companies in times of upheaval: Challenges—business models—management.* SpringerGabler.
21. Proff, H., Brand, M., & Schramm, D. (Eds.). (2024). *Age-appropriate driver assistance systems: Technical, psychological and business aspects.* Springer Gabler.
22. Shaheen, S. A., Chan, N. D., & Gaynor, T. (2016). Casual carpooling in the San Francisco Bay Area: Understanding user characteristics, behaviours, and motivations. *Transport policy, 51,* 165–173.
23. Storme, T., de Vos, J., De Paepe, L., & Witlox, F. (2020). Limitations to the car-substitution effect of MaaS. Findings from a Belgian pilot study. *Transportation Research Part A Policy and Practice, 131,* 196–205. https://doi.org/10.1016/j.tra.2019.09.032
24. Weyer, J. (Ed.). (2025). *Sustainable mobility: How the Ruhr region can achieve the transport transition.* Springer VS.

Outlook: Future of the Automotive Industry

8

Harald V. Proff and Heike Proff

Abstract

This chapter offers an outlook of a book that looks at the future of the automotive industry as it moves from system integration to the ecosystem play. We have analyzed the question of where the traditional automotive companies will remain in the future ecosystem game, i.e., in the race with competitors, tech players, and mobility providers in ecosystems. Thereby, we have endeavored to lay out lines in the complexity of the transformation of this industry, which is simultaneously a skills race for autonomous and software-defined vehicles and in a competitive battle for bundles of offerings with integration of private car ownership and MaaS into comprehensive mobility offerings.

This book looks at the future of the automotive industry as it moves from system integration to the ecosystem game of aligning with a fixed group of partners to achieve an overarching value proposition through multilateral cooperation [1]. We have analyzed the question of where the traditional automotive companies will remain in the future ecosystem game, i.e., in the race with competitors, tech players, and mobility providers in ecosystems. Thereby, we have endeavored to lay out lines in the complexity of the transformation of this industry, which is simultaneously a skills race for autonomous and SDVs and in a competitive battle for bundles of offerings with integration of private car ownership and MaaS into comprehensive mobility offerings.

H. V. Proff (✉)
Deloitte, Düsseldorf, Germany
e-mail: hproff@deloitte.de

H. Proff
University of Duisburg-Essen, Duisburg/Essen, Germany
e-mail: heike.proff@uni-due.de

H. Proff et al. (eds.), *Future of the Automotive Industry*, Future of Business and Finance, https://doi.org/10.1007/978-3-032-26308-7_8

In the content chapters, we initially focused on transaction and innovation ecosystems from an economic (Chap. 2) and technical (Chap. 3) perspective. On this basis, we first looked at globally connected and highly automated driving innovation ecosystems (Chap. 4) and specifically the development of homologation and opportunities for business building in the automotive industry (Chap. 5), followed by an analysis of global MaaS transaction ecosystems (Chap. 6) and specifically of the future coexistence (and cooperation) of MaaS and private car ownership (Chap. 7).

Chapter 2 ("Increasing coopetition between automotive companies in innovation and transaction ecosystems" by A. Jeppe, L. Husemann and H. Proff) has looked at the increasing coopetition (blend of cooperation and competition, when two companies that are competitors in some areas choose to collaborate in others for mutual benefit) of automotive companies from an economic perspective. It has shown, that joint value creation takes place in two very different types of ecosystems: In innovation and transaction ecosystems with differences in the overarching value propositions to the customers, the operating model of actors (between whom data and financial streams flow), the technology and functionality of the technical platforms as well as in governance mechanisms and value drivers.

Chapter 3 ("Technical architecture of ecosystems: Customer application, middleware and backend" by A. Arzani, M. Handte, and P.J. Marrón) has looked at the architecture of innovation and transaction ecosystems from a technical perspective. It has shown the technical interplay of software and hardware on platforms and ecosystems as a base for both scaling and innovating.

Chapter 4 ("SDV innovation ecosystems – An overview" by E. Pritsch, E. Thorwirth, and C Cruceru-Weisbrod) gave an overview about the major elements of the SDV ecosystem. Requirements for the onboard and offboard vehicle architecture in software and hardware were presented. The importance of scaling, collaboration, and continuous development to achieve technologically and economically viable solutions were presented as well.

A closer look has taken Chap. 5 ("Future challenges of homologation and opportunities for business building in transition to SDV as well as the role of traditional OEMs" by E. Pritsch, E. Thorwirth, and C. Cruceru-Weisbrod). Future challenges of homologation in transition to SDVs exist on the one hand due to the insurance relevance of SDVs and on the other hand as a prerequisite for accelerating the market launch of SDVs. The requirements for the technical architecture in SDV ecosystems to guarantee permanent digital homologation were presented as well as the differentiating SDV architecture aspects compared to smartphone and computer ecosystems. Beside the technical updates lots of opportunities for business building in transition to SDVs are possible. The possible future role of traditional automotive companies in transition to SDVs is a move from passive gatekeeper to active enabler, orchestrator, and guardian of the ecosystem.

In Chap. 6 ("Mobility as a Service transaction ecosystems worldwide" by A. Jeppe, A. Arzani, M. Handte, N. Zauner, P. J. Marrón, F. Langanki, and H. Proff) we have used a NLI-based classification method as an innovative NLP application to identify patterns in important MaaS ecosystems since 2015 to find hints for improving the MaaS ecosystems, which have had little success to date. This made it possible to derive the implications that there must first be more "traffic" on the platforms of the MaaS ecosystems, through a more individualized value proposition with additional benefits as well as through more interaction and value co-creation with the customers. Furthermore, the orchestrator of the MaaS ecosystem must not be a pure broker, but has to contribute assets in order to better understand the problems with data flows and governance. After all, PPP are needed to bring together the necessary finances and mobility services and thus to continuously create the necessary framework conditions.

Chapter 7 ("Future coexistence of Mobility as a Service and private car ownership" by H.V. Proff, N. Zauner, and H. Proff) finally highlighted the current coexistence between private car ownership and MaaS and showed that this coexistence will continue, meaning that traditional car companies will not lose their main markets. The demand impulse (market pull), especially from younger and older customers, and supply impulses (market push) through a changed product offering leading to a different market segmentation that includes vehicle leasing, subscriptions, rental, and sharing in addition to privately owned cars, however, require more than a parallel coexistence of private cars and MaaS. They rather require coopetition in a comprehensive mobility transaction ecosystem.

These deep dives on the two main industry changes toward automated and SDV as well as MaaS made it obvious that system integration will no longer be a successful model in the automotive industry. It will become less important with new competitors entering the markets. These are tech players on the one hand, brain suppliers and sharing providers on the other hand, and in addition new competitors from the United States and China. However, the arguments presented in this book make it clear that there is, for three key reasons, by no means an inevitable development that will render traditional automotive companies obsolete.

1. They are not in a situation comparable to that of Kodak during the rise of digital photography (able to become a late follower on the new tech side through digital photography instead of films, but unable to move the business model from printed photography to posted pictures in social media, like the iPhone via additional innovation combining phones and cameras). Traditional car manufacturers may be late adopters of new technologies such as electric vehicles, but there is no iPhone-like disruption in sight, as a car is a far more complex system, operating at high speeds in public spaces, and not "just" a smartphone.[1] This characteristic leads to strong regulation of the cars and also the existence of different customer preferences having different sets of expectations on the many features of the complex product "car."
2. The ecosystem play in the automotive industry is not a true tech ecosystem play with limited physical assets, but an asset-heavy ecosystem play both on SDVs and MaaS. The rules of data-driven tech ecosystems differ fundamentally from asset—heavy ecosystems – sunk costs, physical supply chains, and real-world resource dependencies must be carefully taken into account. Even when the taskforce mindset of working urgently must be overcome toward a

more agile way to work operational agility is different here. It is more a kind of hybrid agile taking into account that the industry is still asset heavy and investment cycles must be considered. Ambidexterity is a given as the value of large investment must be earned to stay profitable.

3. Automotive companies must engage in ecosystems that extend beyond a single industry sector – such as the tech industry – but are instead built on sector coupling. This means merging different business DNAs, legacies, and behavioral logics, which is a complex and often challenging task. Concerning SDVs, this coupling occurs between traditional industrial manufacturing and the tech sector. Concerning MaaS, it spans across industrial manufacturing, the tech sector, and the public sector. These cross-sector ecosystems are significantly more difficult to build and coordinate. Trust is crucial here and this is actually the biggest barrier for a quicker transition both on the SDV and on the MaaS side. New forms of collaboration between the different industry sectors are needed as well.

However, it is not yet clear, who will be the orchestrator in the emerging automotive ecosystems. It could be the OEMs, but it could also be the tech players or the sharing companies who, as dominant players, can determine the distribution of value creation in an industry. Given the significant disparity in market valuations between tech companies and traditional automotive players, it's no surprise that the question of who will dominate emerging ecosystems is often answered in favor of the tech sector – largely due to its superior financial firepower.

Not only since the increasing intensity of competition from tech players or Chinese providers have analysts been punishing traditional, multinational automotive companies with a discount on market value or a low price/earnings ratio [2]. From the perspective of the capital markets, automotive companies are simply not reacting dynamically enough to changes – both in SDVs and MaaS.

Chapters 4, 5, 6, 7 have demonstrated that traditional multinational automotive companies still lack convincing strategies for the transition to autonomous, SDVs and MaaS. Investors currently place greater confidence in innovative newcomers like Tesla, despite their ongoing challenges [2]. These companies are perceived as more future-oriented and capable of surprising the capital markets with breakthrough technologies and products. In contrast, traditional manufacturers seem to lack the visionary appeal that resonates with investors. This is reflected in the EBITDA multiple – a valuation metric representing the ratio between a company's enterprise value and its EBITDA. In the automotive sector, this multiple is only about one-third of that seen in the life sciences industry. In practical terms, this means that an automotive company must generate three times the EBITDA to achieve the same market valuation as a life sciences company, due to the market's more cautious outlook [2].

Figure 8.1 illustrates the development of valuation multiples in the automotive industry over the past five years, with a particular focus on manufacturers of powertrain components, whose multiples have now dropped to very low levels – around 4 – making these businesses increasingly unattractive for investors. Despite all efforts to advance the CASE-transformation, traditional car companies have shown little progress when comparing data from March 2020 to March 2025. Meanwhile, suppliers specializing in conventional powertrain technologies have experienced a steady decline in valuation throughout this period [3]

The low capital market valuation restricts the financial flexibility of traditional automotive companies and, at the same time, strengthens the case for their involvement in broader ecosystems. OEMs often lack the financial resources to acquire emerging tech companies – therefore, cooperation becomes a necessity rather than a choice.

All companies are actively working to enhance their own valuation, with the equity story playing a pivotal role, as it helps reduce perceived risks for investors [2]. A compelling equity story typically consists of four key elements: (1) market development, (2) strategic positioning and business models, (3) core competencies, and 4. corporate governance.

1. Market development in the traditional automotive business (covering vehicle development, production, sales, and aftersales) is increasingly unattractive for automotive companies, as illustrated in Sect. 7.2 in Chap. 7. Unit sales growth has largely stagnated. To demonstrate growth, companies must either gain market share from competitors – which, in the highly competitive automotive industry, is likely to come at the expense of profitability – or generate more revenue per vehicle over its (extended) lifecycle. This can be achieved through data-driven services, software updates or multi-life models for key components such as the battery (see Chaps. 4, 5, 6, 7 for details). As a result, SDVs and MaaS ecosystems must quickly become core revenue and profit drivers for OEMs. After all: without a fully functional SDV and MaaS ecosystem, hopes for a better market valuation will remain unfulfilled.

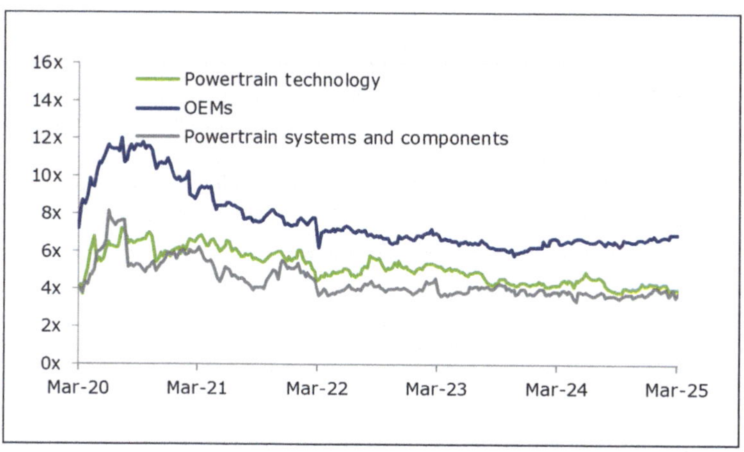

Fig. 8.1 Development of multipliers in the automotive industry. *Source* Deloitte [3]

2. The shift from traditional business models – characterized by hierarchical, linear value chains and optimized make-or-buy decisions based on high bargaining power – toward business models as orchestrators within complex ecosystems of equal partners presents a significant challenge for incumbent automotive manufacturers. These companies often struggle with the principles of coopetition (see Chap. 2), as their legacy structures are not well-suited to collaborative and competitive dynamics within ecosystems. As a response, many automotive companies are establishing new business units specifically designed to navigate the ecosystem-oriented paradigm, thereby creating institutional separation from their traditional operations. This allows them to maintain relationships with conventional suppliers while simultaneously engaging with new types of partners. Notably, major Tier 1 suppliers are themselves increasingly positioning to act as partners within these emerging ecosystems.
3. In the transition toward ecosystem-based business models, automotive companies face a significant gap in two key areas of competences. First, competences in software development are crucial, especially for the deployment of SDVS and the development of MaaS offerings. In response to this need, some traditional automotive manufacturers have now employed more software developers than certain technology firms. However, this increase in software talent does not necessarily transform automotive companies into agile companies capable of working with a "minimum viable solution" approach, which is typical in the tech industry. Secondly – countries like Germany, Japan, Korea, and with this the companies with headquarters in these countries are proud of their mechanical engineering champions league capabilities. But they are not at this level in software engineering because the universities and the tech companies out of these countries are on average not there yet. This must change quickly otherwise these locations will no longer be important for global companies. Thirdly, the development of ecosystem management competences – such as partner management and the coordination of collaborative activities – remains a major challenge for many automotive companies (see Chap. 2). This is an area in which many traditional firms continue to struggle. This missing hybrid agile DNA needs to be built up if the stock market is to seek a different valuation for automotive companies.
4. The transition to ecosystem-based business models has yet to result in significant changes to the governance structures of automotive companies that would prompt analysts to revise their equity stories for these firms. This must begin at the top management level, as this is where the most significant strategic shifts need to occur. While there have already been numerous transitions involving the appointment of tech executives to key positions in automotive companies, these efforts have not been as successful as planned. This underscores the necessity for additional substantial and strategic changes in leadership and organizational structure.

Without these deep structural adjustments, traditional automotive companies will struggle to navigate the rapidly evolving technological landscape, particularly in areas such as autonomous and SDVs, and MaaS. This discussion highlights that the automotive industry's hope of achieving a similarly favorable market valuation as that of technology companies through the shift from system integration to ecosystem play has not yet been realized. This is precisely why the capital markets are currently anticipating more positive surprises from California-based tech companies, with the "one more thing" that could spark the next technological revolution. In contrast, traditional automotive companies present a much more predictable trajectory, as they tend to delay their plans, thereby dampening investors' imagination. It is a perfect moment for surprises via "out of the box" solutions overcoming existing conflicts like the either/or approach in mobility (see Chap. 7).

The competition for the most advantageous positions within the emerging ecosystems of the automotive industry is far from decided, as individual mobility remains a complex, capital-intensive, and highly regulated market. However, one thing is clear: the race for the future of the automotive industry is underway, and inactivity is not an option. The chapters in this book have outlined potential solutions, yet there is no universal remedy or single "world champion" strategy. Consequently, each OEM and Tier 1 supplier must accelerate its own path from system integration to ecosystem play, tailored to its unique competencies.

Notes

1. Prof. John Paul MacDuffie, Professor of Management and Director of the Program on Vehicle and Mobility Innovation, The Wharton School, Philadephia, USA, during a panel discussion at the 32nd Gerpisa International Colloquium in Bordeaux, 2024.

References

1. Adner, R. (2017). Ecosystem as structure: An actionable construct for strategy. *Journal of Management, 43*(1), 39–58.
2. Proff, H., & Proff, H. V. (2013). Dynamisches Automobilmanagement: Strategien für international tätige Automobilunternehmen im Übergang in die Elektromobilität (2. Aufl.). Springer Gabler.
3. Deloitte. (2025). *Internal capital market review monitor*.

The manufacturer's authorised representative in the EU is Springer Nature Customer Service Centre GmbH, Europaplatz 3, 69115 Heidelberg, Germany. If you have any concerns regarding our products, please contact ProductSafety@springernature.com

Printed and bound by CPI Group (UK) Ltd, Croydon, CR0 4YY
25/09/2026
02222836-0017